Bringing the Mountain Home

THE UNIVERSITY OF ARIZONA PRESS / TUCSON

Bringing the Mountain Home

SueEllen Campbell

The University of Arizona Press

♾ This book is printed on acid-free, archival-quality paper.
Manufactured in the United States of America
01 00 99 98 97 96 6 5 4 3 2 1
Library of Congress Cataloging-in-Publication Data
Campbell, SueEllen.
Bringing the mountain home / SueEllen Campbell.
p. cm.
ISBN 0-8165-1616-2 cloth (acid-free paper). —
ISBN 0-8165-1617-0 paper (acid-free paper)
1. Natural history. 2. Nature. I. Title.
QH81.C3525 1996 96-4466
508—dc20 CIP
British Library Cataloguing-in-Publication Data
A catalogue record for this book is available from the British Library.

For Audrey Dorsett,
who taught me to feel at home
in wild places,
and for John

Contents

Preface

Let me start with a story about a walk high in the mountains. It was September, I think, when snow flurries had replaced daily lightning storms but hadn't yet closed the trails above timberline. The sun was warm, the sky brilliant, the wind icy, the sturdy low flowers of the tundra gone to seed, dwarf willow smoldering russet.

Bundled in sweaters, mittens, and earmuffs, I wandered up the faint, familiar trail to the ridgeline, my attention drifting in and out of the moment. Daydreaming, I looked for signs of elk, bighorn, marmot, ptarmigan. Thinking about the week just past, I leaned over to feel the rasp of granite and wrinkle of lichen. Every few steps I paused to absorb the view—the Mummy Range filling the space in front of me, the Never Summer and Snowy Ranges emerging behind me and to my left, the big indigo peaks of Colorado's Front Range stretching far past the horizon on my right—and thought about the book I'd been reading, something about high mountains by John Muir. Like his landscape, I reminded myself, mine had been shaped by glaciers.

I pulled out my canteen for a sip of water and noticed what I usually don't—its cool, smooth, shiny red surface, scratched and dented by nearly thirty years of use, its faintly metallic and intimate flavor. Muir, said the voice inside my head, didn't carry a canteen; he was far more ascetic than I, probably took no notice of his thirst, and didn't have to worry about nasty wild-water bugs like giardia. Surely, though, he must have had moments very much like this one—and, the voice continued, like most of the other moments I've spent in such wild places. And so must have countless other people, perhaps thousands on this hillside alone. Many years before I was born, my father drank from his own familiar canteen on this trail, and my teenaged niece was here just last month.

In fact, I realized, I was taking two walks at once. One was intensely personal and immediate, *my* body, senses, memories moving through a specific and extraordinary place and moment. The other was shared, my own experience formed by my culture, by other, earlier, visitors to wild places, by circumstances, attitudes, assumptions, words, even emotions I had no part in creating but had somehow absorbed into myself.

On the first walk, I was alone because my husband, John, who hates to let me go first and hikes faster than I do, had moved far enough ahead that I could feel the solitude. On the second walk, I was *alone* because I'd been reading Muir, because my culture has so often portrayed wilderness travel as a solitary experience, one person moving through an empty landscape, all companions written out of the record. On the first, I was out of breath and a little tired because I was at twelve thousand feet and, as usual, a bit out of shape. On the second, I was doing some mild *trudging*. When I stopped later for lunch on the first walk, a marmot, amazingly, inched toward me until he stood right on the toe of my boot, so brave, perhaps, because he lived in a national park where other hikers might have fed him tidbits. Lunchtime on the second brought an *encounter with a wild animal* in a *pristine wilderness*.

Another familiar book, I remembered, could give me a framework for thinking about this doubling. In *A Lover's Discourse*, Roland Barthes borrows the word "figure" from choreography, where it means something like a momentary attitude of the body or a fragment of a dance. He explores how the language and experience of a lover are composed of such attitudes, such fragments—each affair at once an original, one lover, one beloved, one day's intimate event, and at the same time a collage of shared moments, feeling engulfed, waiting for the phone to ring, making a scene. The same thing, I realized, could be said about the stories and experiences of wilderness travelers.

I started seeing these figures everywhere—through the rest of that September day, in each trip I took over the next few years to a wild place, in every wilderness narrative I read or recalled. I made lists, played with names, tried different combinations and distinctions. I identified concepts like pristine, actions like basking in the sun, emotions like stunned by beauty. Sitting on my living room floor packing for a camping trip, I'd think: chaos, lists, traveling light, gear, food, obsession. Running down a steep scree slope, deafened by the crack of thunder close behind me, I told myself, "If I survive, this will make a good story. Fear? A brush with death? Exposed? Just plain lightning?" Every time I caught myself yearning for another trip to a wild place, I'd see that I was immersed in desire.

I knew how they *felt*, these momentary attitudes of my body, but now I also wanted to think about what they might *mean*. What fine webbing, I wondered, holds ideas to bodies, words and stories and cultural conventions to the sensory immediacy of licheny rocks, the taste of an old canteen, the gusts of memory and desire? On that vivid September day in the tundra, what made those shared wilderness figures seem so thoroughly my own?

So I started this book. I decided not to write about figures I know mainly from reading (being alone or lost or frozen halfway up a precipice, at-one-ness) or from the stories of friends and family (getting a workout, the allure of fly-fishing, the way figures like quiet and bugs twist and re-form when you take small children on camping trips). Instead, I've explored some of those I've found most fascinating, insistent, seductive, the ones I've learned from the inside, with my bones, my muscles and nerves, my heart.

These are the figures that keep luring me outside, that map my own landscapes of desire. I've shared them with many others, I know, and I've never been the original choreographer. Still, the dance they trace, this dance in wild places, is always my own.

Bringing the Mountain Home

Desire

Pressed flowers slip from the pages of my book. White fairy trumpets from the basin below Silver Mountain—a rocky hillside of small abandoned mine holes, piles of saffron and rust smelling lightly of metal—where their usual bright scarlet bleaches to the faintest of pinks. Yellow glacier lilies from high in Hummingbird Basin, snug against the edge of melting snow in late July. Strong enough to punch through inches of winter, they're the most elegant of wildflowers, curved and swaying from fine arched stems. Fringed gentian, rich blues held like a candle in papery green, from one of the wet meadows at camp.

They might be fifteen years old, or thirty. I would have gathered them carefully, each from a patch of color thick enough to spare a few potential seeds. Kneeling on the damp or rocky ground, my back to the lightest breeze, I'd pull a thin paperback from my pack—René Daumal's *Mount Analogue*, Han Shan's Cold Mountain poems—and spread a Kleenex over the fold. It would take ten fingers, the heels of both hands, and concentration to arrange each flower, think how to preserve the floating grace of three dimensions in just two, crush the blossom's thickness gently against the page, stroke each fuzzy leaf into place. Some flowers would flatten easily, their centers filled with shadow and sunlight. Others would stay dense and fleshy. Right hand spread across petals and leaves, I'd cover everything slowly with the rest of the Kleenex, close the book, wrap it with a rubber band, and put it back into my pack. Home in my cabin at camp, I'd slide each tissue across to a fat book, one with hundreds of pages and maybe even hard covers, *Finnegans Wake* or *Don Quixote* or *Anna Karenina*, some complex tale of dreams, quests, desire.

There they've stayed, many of them, fading, translucent, and fragile, waiting to startle me with memories and images. Sometimes, though, I'd pull them out when they were completely dry and make note cards with them. A layer of waxed paper, plain side down, the flower, a sheet of white Kleenex, painted painstakingly together with a small brush and mixed glue and water, left to dry, then ironed into parchment and folded over writing paper. After one snowy winter, the glacier lilies were so abundant that I made a whole set of cards for my mother, who surprised me by saving them to use for condolence notes. Now I understand why—how tightly knotted are beauty and transience, loss and desire, the heart-catching flower and the fragile stem.

Wild flowers, from wild places. As familiar to me as home. I gathered them in the summers of my teens and twenties, during the months I spent as a camper and then as a counselor at a small camp on a cattle ranch in the Colorado Rockies. It was called Skyland. My parents worked there before I was born, the year it opened, and promised me that when I was twelve I could go. I returned nearly every year until my late twenties, through junior high and high school, college and graduate school. My three brothers went to the boys' camp and I to the girls', a quarter-mile up the hill, past the barn and corral and ranch house, on the shores of a small lake. We all loved it.

It was a very beautiful place. Often in the evenings I'd sit on the wooden step of my cabin to comb my hair dry, my feet tender from a day's hiking, the skin on my face warm and taut with sunburn. I'd squint a bit in the low sun and look at the view. In the grass below me gleamed three silver canoes, and two rowboats nudged a small dock tucked into the green and yellow gloss of water lilies. My body knew the painful iciness of that water, only a little milder on the sunlit surface and above the scattered warm springs; the waxy thickness of the lily pads, how slippery they were underneath; the slap and pull of paddling against a sudden wind, into sharp waves. Sometimes in the mornings clouds would hang low over the valley and cling to the lake, turning it soft and otherworldly.

To the right, steep hills climbed through spruce and boulder fields to pale crumbling cliffs. Every few days, at noon or twilight or as I waited to fall asleep, I'd hear the rumble and clatter of a rockslide somewhere over my head. Lower, rounder hills to the north were patched with aspens, soft green puffs from this distance, their leaves always shifting, glittering, swishing and clattering faintly in moving air. On their pale silky bark was a fine powder I liked to smooth into the planes under my cheekbones.

To the west, grasses and sagebrush covered the hills, in shadow now in the evening, but a wonderful silvery blue green in the sun or in moonlight. When I walked there I'd pick a handful of sage leaves and rub them over my wrists, the inside of my elbows, behind my ears, over my throat. Sometimes the evening air would carry a faint hint of this perfume, dusty with sun or moist and intense with rain, and I'd remember the coyote I'd seen early one morning. I'd ridden out to bring the horses in from the far pasture where the fringed gentian grew, and she materialized, silver fur in silver sage, and walked along with me for a while.

Behind all this tall mountains circled the valley in shifting shades of red, green, white, and blue, deepening in shadow, washed at sunset with ground jewels, ruby and sapphire, emerald and gold. They'd still be thick with snow in June, and on late August mornings we'd wake to find new dustings of white. In the evenings, gazing at their soft reflection in still water, I'd think about their high scree slopes and casual waterfalls, their sweet shoulder-high flowers and browsing deer, the feel of hot sun or sharp hail on my skin, all their secret mountainy details.

During my first summer at Skyland I was often homesick for my parents and my house on the edge of Denver. I remember the sharpness of that pain, how it tightened my throat and lay heavy in my stomach, how it unsettled me. Slowly, over the next summers, I regained my poise. I learned to paddle a canoe, open a barbed-wire gate, cross a field of boulders with sure steps, careen down a snowfield on a folded poncho, saddle a horse and sit a trot. I learned to build a fire, organize a backpack, read the sky for coming weather, put names to wildflowers, follow a faint trail and do without one. I learned how to see, how to move through grandeur into intimacy, how to make myself at home in wild places.

In my late twenties the tourist industry discovered this valley. The rancher sold his land and Skyland lost its lease. Earthmovers chewed up the horse pasture and gentian meadow and spat out a golf course. Bulldozers gashed roads through the aspen groves and across hillsides. Gargantuan condominiums, mansions, and a clubhouse replaced the covered wagons my brothers had slept in and the small cabins with the view of the lake. A long series of consortiums and multinational corporations put their names to the deeds. The ranch house was left to stand as local color.

Desire, some say, arises from loss. As water scooped into a cup is no longer part of the pond but holds its own small mirror of the sky, we become our separate selves only when we lose touch with some original wholeness. Apart from God, from our mothers and fathers, from some other half of ourselves, from home, from the natural world. There's no way back, but desire is the force that keeps us trying. "When I am from him," wrote Sir Thomas Browne over three hundred years ago, "I am dead till I be with him. United souls are not satisfied with embraces, but desire to be truly each other; which being impossible, these desires are infinite, and must proceed without a possibility of satisfaction." Desire shifts like the wind, it gusts and whispers and swirls around us, it pushes and tugs at our hearts. Then it tugs again.

But I also believe the desire for wildness is an elemental force, like gravity, like magnetism. With the power of a lodestone, a charged center of mass and energy, a deeply loved landscape holds us fast to the planet. Drawn to one wild place, to a small lily-splashed lake in the Rockies, I'm drawn to all wild places. To the other landscapes of home and to the far and exotic. To glowing sandstone deserts and liquid dark jungles, to junipers, acacias, and giant mango trees, to savannas and swamps and steppes, to white pelicans, pink flamingos, and emerald parrots with scarlet-streaked wings. Wild beauty, dazzling.

I sit at home and imagine journeys. I'll walk, I think. Maybe I'll go back to Hummingbird Basin or check on the white fairy trumpets beneath Silver Mountain. Or maybe Utah in the spring—Capitol Reef or the Book Cliffs or the San Rafael Swell. The Grand Canyon is always more stunning than I remember. Alaska! I'll get a pilot to fly me somewhere incredibly remote and pristine, maybe the Brooks Range, and drop me off for a month among the grizzlies. If I save enough, I can go to the Serengeti or Madagascar or Tierra del Fuego. Maybe I could see an iceberg or a penguin. In January I could go look for whales off Baja California. I'll float on a river, I'll climb a mountain. I'll get dirty and tired and sore. I'll soak up the sunlight and the space and the silence.

I pull out travel books, nature books. I turn the television to PBS. I leaf through the world atlas and catalogs full of camping equipment. I dig through my pile of topographic maps. I roam around my house touching bits of lichen and mica, sea shells and birds' nests, seeds and pinecones and feathers, river rocks, sprigs of sage. I open an old book and watch wildflowers fall out.

Equipment

Gear is everything, says my friend Ben. Maybe he's right. I'm sure surrounded by it now. We leave tomorrow for Wyoming's Wind River Mountains. Six days, two rented llamas, four people—my husband, John, our friends Mary and Bruce, and me. It's mid-August, the beginning of snow season. I'm in charge of the food for everybody and everything but John's clothes for the two of us.

On the kitchen table in front of me lies a tangle of lists—on graph paper, computer paper, jagged newspaper margins, scrawled big, crossed out, neat tiny lines and sums: what's left to buy and where (cashews, fuel, water filter; grocery store, army surplus, outdoor shop), what needs repacking in smaller containers, what to eat at each meal, exact weights for everything. The bathroom scale is on the living room floor, the kitchen scale at my right elbow.

Our tent weighs a hefty ten pounds, our warm and roomy sleeping bags seven each. Three indispensable pounds for our Therm-a-Rest pads—no more sore hips keeping me awake. The stove plus fuel only two pounds, six ounces; breakfast food two pounds without coffee. Two more for the nest of pots, bowls, cups, spoons, and candle lantern. (Pure sentimentality, that lantern: the moon will be full. But my first sweetheart gave it to me, and though I rarely use it, I take it on every camping trip.) Extra clothes, five and a half. John walks by and rolls his eyes: Easy for him to say, I mutter. Fish, I remember, and fix a tiny plastic bottle of olive oil, a Ziploc bag of cornmeal, a film can of lemon powder.

Jungles of stuff cover the counters—a big box of dry milk, jars of jam, an empty fuel can, Dr. Bronner's Peppermint Soap, a first-aid kit waiting for antiseptic. In the living room, plastic bags stretch and bulge with weird but equally heavy loads. In the guest room are heaps, drifts, boulder fields of miscellany, bandannas, patch kits, spare candles bent like skinny pink cucumbers, rusty sheets of uncut moleskin, stretchless Ace bandages, garter snakes of cord, ancient stuff sacks, never-opened emergency blankets.

It's a familiar scene, really, since I always do the planning and packing for our wilderness trips. Somehow John always has another job to do, a suddenly urgent one; plus, he'll point out if I ask for help, he'll load the car and check the oil and tires. And I mostly enjoy it. It's like playing with dolls—the fascination of miniatures, the allure of life on a smaller scale. It's a challenge of efficiency, too: the lightest gear, least space, everything essential, no excess. I also like the *idea* of traveling light. And—not least—all this preparation expands the trip, adds days to mountain time that's always too short.

But this packing job has me feeling a little obsessive. Weighing a plastic cup and spoon is not normal behavior. Why am I doing this?

One reason is right on the surface, my most obvious excuse. I'm trying to pack light—but not *too* light. Each llama will take just sixty pounds and we don't want to carry much ourselves, but Mary's a newcomer to wilderness travel and she's feeling anxious about being so far from roads and phones. I want her to love this trip—I need more companions for hiking and camping. So I'm trying to soften the wildness a little, spin Mary a thicker cocoon of comfort, with equipment for glue and food for silk.

The monologue in my head reminds me there's a second reason: I've lost my packing innocence. I've just finished a long research and writing project about wilderness food, how it tells us things about its eaters, and I can't stop thinking about the *significance* of every item on my menus. Every bite is an idea, I recite to myself, every calorie a concept, every menu a dissertation. I decided on our meals days ago, but I keep rehearsing the options.

Minimalism, for instance. On the model of John Muir, who took little more into the mountains than tea, dried bread, a tin cup, and an overcoat. (Or so he said. I admit I'm skeptical.) Spending a long time in a pristine wilderness, leaving no trace, consuming no other lives to fuel our bodies: the symbolic implications, however unrealistic, are appealing. But hunger isn't.

Then there's "living off the land." Maybe we could just eat fresh fish and scrounge a few plants? (I have a book with pictures.) To test the idea, I mentioned that we might count on trout for one meal—after all, the Wind Rivers are famed for their fishing and Bruce and Mary are pros. They just laughed. "Sausage and cheese," Bruce said.

Or the other extreme, gourmet. With the llamas, themselves an elegant touch, we could carry a bit of extra weight: cabernet sauvignon? Brie? some pricey freeze-dried treat like shrimp au gratin or chocolate fondue? Then we'd be right in the tradition of the so-called gentlemen hunters of the nineteenth century, who lugged along wagons full of wine and silver and stocked their extravagant tables by slaughtering the wildlife. Gourmet camping is still big business, I know, but I want none of it. Too much city, too little wildness.

Ordinary camping food offers a conceptual middle ground. There's a tradition here, too, or two of them—military rations, with "conquer the wilderness" implications I can do without, and home cooking, which suggests the much more attractive "make a home in the wild." But there are practical difficulties. Lots of classics are too heavy for the llamas—beans and bacon, biscuit mix and a dutch oven, pancakes and syrup, canned tomatoes and peaches. We can't cook anything wrapped in foil and buried in a roaring campfire: where we're going, open fires are forbidden. I don't want to waste stove fuel, either, and cooking is a lot slower at high altitudes. We'll stick with fast foods. Ramen, maybe, or wonton soup? In my cabinet lurk petrified samples, souvenirs of my brother Bruce's backpacking days. It seems a waste to throw them away, but I sure don't want to eat them.

So my menus are hybrids, fancier than usual for Mary's sake. We'll take one serious anxiety-calming luxury, we all agree, a bottle of single-malt scotch (sacrilegiously transferred to a plastic canteen), and I've packed olive oil, fresh garlic, and several other gourmet touches reaching their metaphoric peak in a tube of pesto and some sun-dried tomatoes. Our carbohydrate staples are lightweight and fast-cooking: angel hair pasta, dried black beans and minute rice, couscous. Contemporary camping food classics. For the final night (it will be my most popular creation) we'll have the couscous cooked with a can of chicken, cashews sauteed with garlic in olive oil, a pinch of dried red pepper flakes, and a couple of packets of soy sauce saved from some forgotten fast-food place. Eco-gourmet. My food-as-symbol conscience is reasonably clean.

Still, there seems to be something more behind these chaotic piles, this library of lists, this obsessive planning. One more reason. Finally I identify it. With kitchen scales and Ziploc bags I'm trying to enchant our equipment, cast a benevolent spell that will make it serve us silently and well. I'm haunted by my last trip, a week in May car-camping with another friend in Chaco Canyon, when mutinous equipment almost took control.

I'd finished reading my students' final papers on the plane to Tucson and phoned in the grades from Nina's apartment. Far too hot, rushed, frazzled, we'd thrown our gear together in a frenzy, put gas in the car, ice in the cooler, car-coffee and road-doughnuts in ourselves, and left town—talking as fast as we drove, catching up on the last ten months, determined not to waste an hour. We thought that we had everything we'd need (after all, I'd made lists, I had experience), that our rush didn't mean we weren't paying attention. We thought we were really *ready* for this trip.

We were wrong. By the time we arrived, the campground was full and we had to set up our tent on a hard dirt side road. We had no hammer for the stakes, so we tied one end to the car bumper and the other to a pair of water jugs. It was after dark and moonless, there was no nearby electric light, and our flashlight was lost somewhere in the trunk. In the morning, we set up our borrowed Coleman stove, put water in a pot for coffee, dug out the box of kitchen matches. It was almost empty. The first three matches wouldn't strike, and when one finally caught, the stove flamed for a few seconds and then died. We tinkered and tried again, then swallowed our pride and asked a competent-looking guy from the next car if he could get it to work. He couldn't.

We moved into the regular campground—flush toilets, running water, other campers every few feet. The ground was in turns too hard and too sandy to hold the tent stakes. Nina's new sunshade blossomed into a wind sail. A pack of drummers drummed past midnight—male bonding? imagining the ancient hunt?—and a dozen exuberant high-school students shrieked past our tent every time we dozed off. In the restroom, Nina listened to one girl tell another that she shaved her armpits every single morning. All day long sand blew into our tent, and one night a gritty valve kept my sleeping pad deflating, though I blew new air into it every time my sore hips and shoulders woke me. These were not the details we wanted to be paying attention to. We'd come because we wanted to hear the whispers of Chaco's mythic past, but there was so much noise around us. Only when we'd walked for hours, miles away from our camp, could we begin to listen.

We never did get the stove to work properly, though we took a whole day to drive out to Farmington and buy a part, a new generator. Hefty tribute, we thought, a generous offering to the spirit of equipment. But it wasn't enough. With the perversity of its species, the stove lit perfectly in the hardware store's parking lot (we whooped, congratulated ourselves on our feminist self-sufficiency), but back in the campground it reverted to rebellion, working only, it seemed, when the sun shone on it. So we tinkered, begged, cursed, and finally adapted ourselves, shrugged off the annoyance, cooked extra when the stove lit, ate cold food when it didn't. That machine might not be our servant, but we'd try not to be its slaves. I did figure out—with a surprisingly deep satisfaction—how to clean the sand from my sleeping pad valve. And we coaxed the sunshade into acting as a rain fly for the tent—just in time for a storm. As we left, we gave our extra food to a pair of hungry sisters in the next campsite. We were thrilled to find anyone less prepared than ourselves.

Equipment almost devoured this trip, though we struggled first to control it, then to ignore it. After all, we kept telling each other, the important point was to live with less, to lighten and simplify our lives (Thoreau's a permanent voice in my head), to pay attention not to housework (enough of that at home) but to the *real* world around us. The Chaco people, we said like a litany, didn't even use wheels—an almost unimaginable difference. *They* made their meals without Coleman stoves or matches—or ice chests or a visitors' center with hot water.

So this time I'm determined to do what I can to keep the same fate from overtaking us in the Wind Rivers. I'll pay so much attention to equipment before we leave that I'll be able to ignore it completely in the mountains, and then everything will go perfectly. Gear, I've learned, demands its due. The little leather cup that makes the backpack stove work has been soaking in oil for two days now (the sad lesson of another fireless trip), and I've packed matches in four places, one of them a whole big box full, two of them waterproof, sprayed Scotchgard over my increasingly leaky poncho and slathered Sno-seal over our hiking boots, written cooking directions on each Ziploc of dinner food. *This* time I'm *really* ready.

Two days into the Wind Rivers and I'm foiled again. Oh, everything is working, but I feel like Atlas, the weight of the trip on my head. At first only I know how to use the water filter; John's forgotten how to work our stove; it seems far easier to do all the cooking than to explain the menu and recipes to anyone else. While the other three wander off to explore the lake below me and climb the hill beyond, I stay in camp and see to the gear—take the stove apart to pack it, reorder the lunch food, stuff sleeping bags. Again my focus is too short, my wilderness world the gaudy drifts of nylon and fleece and plastic and aluminum surrounding me.

Have I simply run into a fundamental reality, what Colin Fletcher calls the "stubborn and inescapable paradox of simple living"? Even in wild places, is there no way around the distraction of equipment—and what it stands for, keeping the body fed, clothed, sheltered—other than cheating and leaving the job for someone else to do? Practically every wilderness writer I've ever read talks enough about equipment to suggest that this might be so—men at least as much as women, reassuring me that I'm not just suffering from deep gender training. Or is it just that my journeys are never long enough for these trivial matters to become wholly routine, wholly mindless?

I sit washing my breakfast bowl, wondering why I can't manage to escape such time-wasting chores, then, gradually, why I should so much resent them. These are pleasant minutes, really, in the cool air and warm sun of the morning. The icy water feels good on my fingers, as does the small spruce brush I'm using as a scrubber, and two tiny chipmunks are putting on a show, careening around the tent and over the boulder on my left.

Could it be, I realize all at once, that I just need to think differently? Thoreau's "Simplify, simplify" is joined by the words of the Buddhist teacher Thich Nhat Hanh. There are two ways to wash dishes, he says. One is in order to have clean dishes. When we do them this way, we are not really alive, we are putting off our life to another time, ignoring and wasting the present moment. The other is in order to wash the dishes. This way, he writes, "The fact that I am standing there and washing these bowls is a wondrous reality. I'm being completely myself, following my breath, conscious of my presence, and conscious of my thoughts and actions." In the moment, mindful, paying attention.

In Chaco, I wonder, did Nina and I somehow *need* the astringency of that broken stove, those forgotten matches, to draw in our attention? Before we could begin to imagine ancient lives, did we have to settle into our own? Maybe my fascination with the detail of the Chaco people's days, with the technology that framed their culture, grew from the unstable roof of our tent, the insistent trouble of cooking our meals. How did they manage in this hot and spare place? Like us, they needed to cook, eat, keep clean, build houses. Like them, we were—we *are*—surrounded in every moment by beauty and spirit.

Maybe the real privilege of a week in Chaco Canyon was to make a kind of home there, with whatever gear we had. Simply to prepare an ordinary meal with right attention—on those warm yellow rocks, in that wide dry canyon, under that transparent blue sky, in the company of lizards, ants, spiders, ravens, a good friend, the whispering past. And this icy water, fragrant spruce, warm sun, speedy chipmunks, clean bowl—maybe this moment is the point, the destination, of my Wind River journey.

Packing for my *next* trip, I'll try to remember.

Finding the Way

This trail is wilderness neon. It's a large-print book, a billboard, giant arrows seared into the ground by visiting space creatures. Don't get lost! it shouts. Come *this* way.

We pass carved and yellow-painted wooden signs on chest-high posts at junctions: SURPRISE POND →, SOMEWHERE ELSE ←. Sister signs at curves in between: TRAIL ↑. Blazed trees every six strides, sap-stained slashes in bark, white paint, red paint, layers of trail ink. Ribbons of yellow plastic dangle from scattered branches: Here's the trail! A present!

We're walking on the Pingree Park campus in northern Colorado—summer home to decades of trail-besotted forestry and recreation-resource students. We carry a hand-drawn local map. "Caution," it warns: "Trails in this area are difficult to follow due to the terrain. Pay attention!"

But this path is four feet wide, more dirt than rock, scrubbed clean and rinsed lightly in pine-needle rivulets. Dead trees have been sawed off flat and cleared to the sides. Their bright pale disks mark the edge like lines of highway reflectors, like city streetlamps glowing at twilight.

John walks ahead of me, stops, bends over and draws an arrow in the path, pivots to face me, spreads his arms, waves me toward him as if he were ushering an airplane to a concourse gate. With his muscular shoulders and dark, curly beard, his plaid flannel shirt, Levi's, and ancient Yankees cap, he looks like a lumberman, a short Paul Bunyan, though he's really, like me, a teacher and writer. We laugh at his joke, walk on. He stops again, stoops, builds a tiny cairn of fist-sized rocks, then semaphores a curve in the trail.

How could anyone possibly lose this way?

I play one of my wilderness mind games: Erase the Trail. Vaporize the too-smooth stumps. Heal the blazes, wash off the paint, collect the plastic bows. Suture the cut trees back together where they fell, or move them away and wait for others to grow tall, die, topple in high winds. Shovel pine needles and bark chips over the dirt; scatter seeds for tiny plants with lovely silly names—grouse whortleberry, little pipsissewa. Then wait. Watch the bright green and smoky gray lichens recover the worn granite.

At what moment, with what healing gesture, what unfurling leaf or falling pinecone, would the trail finally disappear?

"The trail veers right," says our map, "and reaches Surprise Pond where and when you least expect it!" We walk slowly, looking for the gleam of hidden water. A big new sign stops us short with giant letters: SURPRISE POND! A pond, indeed, a pretty little thing, destination enough for our walk, one I know we'd never have found without help. What pleasure it must have given its unsuspecting discoverers, its trail builders, its mapmakers! I don't know what to think: with their innocence and enthusiasm, all these directions are amusing, even charming, but they've certainly left no room for surprise.

I find a smooth rock and sit in the sunlight. On the blank pages of a small notebook I sketch the sway and droop of seed-heavy grasses and think about maps and trails.

What if there were no path, no signs, if this place had no name? What if we walked right by without turning our heads, if no one ever climbed this small part of this inconspicuous ridge in this vast glacial basin? This tiny complete landscape, glinting still water, bright reeds and rushes, soft grasses feathery with russet seed, aspens just turning to fall—in what mysterious way would all its loveliness change?

We like to think that in the wilderness we escape streets and signs. We venture beyond familiar places where everything has been named, made human, possessed, where all paths are known, mapped, set in concrete or ink. A wilderness is roadless, both by agreement and by law. Surely this should also mean trail-less, sign-less, mapless, nameless: no trace of human writing on the land, nothing to say that we have inscribed this place as ours. An absence that signals the purity of the land, an absence at the center of our desire.

Maybe we should go to the wilderness to get lost, to lose the familiar way of cities and towns, to let loose of our everyday sense of our place and find another way of being in the world. Lost, *amazed*, I might forget myself and find myself, a creature among other creatures, a reed in the wind, fed by sunlight, dead plants and animals, minerals from the mountains crumbling beneath my feet.

Sometimes I do walk out across country where there are no trails. I leave my maps behind and don't think about the names of creeks, ridges, valleys. In a new place I feel my way slowly, study plants, soil, prints of small animals, scratches and scars on aspen bark, munched ends of willow limbs. I listen for birds, feel the power and angle of the sun and wind on my skin, in my hair. I look for game trails and follow them. I try to read the language of the land, spell out and decipher its signs, etch into the canyon walls of my mind its verbs and nouns, its grammar and syntax. Sometimes in snow the tracks of rabbits trap pools of blue shadow; sometimes they glow with captured sunlight, the warmth of passing creatures held visible. With my eyes and in my mind, I trace rivulets and streams, the vanished routes of glaciers, the hills, basins, canyons carved by wind, water, ice.

On a trail, you don't *have* to pay attention to where you are. Without one you do. With every step, with every detail you absorb of scent and shape, fragments of earth and sky shift and settle around you into patterns of landscape, as if the bright shards in a kaleidoscope were to resolve themselves into a cathedral full of stained glass windows.

Most of the time, though, I like to know in human signs just where I am. Normally I walk on trails. I know the names of creeks and mountains, or I wonder what they are. I read Forest Service signs, I note tree blazes even along obvious paths. I pay attention to the moment I cross the posted boundaries of wilderness areas, as if the steps beyond were in a new world. Sometimes I carry maps even when the trail is short and well-worn, or when my way is as familiar as the sidewalk to my parents' house.

Why? Partly it's just routine: along with a space blanket and matches, we carry maps just in case. Blizzards can blind you in minutes, ankles can sprain in a step, some shortcuts warp time and space, turn ordinary landscapes into Dalí paintings.

Still, I'm rarely anywhere it's hard to see the way, where the sun stays hidden or forests obscure the forms of the land: in the Rockies where I live, in the deserts to the west, long views are everywhere. And I pay close attention to landmarks, landsigns. I've read that our sense of direction—like that of migrating birds—might have something to do with magnetism and traces of iron inside us. Mine is a strong one, my sixth sense, tiny black flecks forming the lines of maps in my brain, shifting steadily as I move, turn, look around, align my own body's magnet with the earth's. I am not afraid of getting lost.

Is it just a quirk, then, my fascination with mapping, with human signs in the wilderness, with finding the way? It seems not. I go for a short walk in the Sierra Nevada with a friend who knows those mountains intimately. We take the Pacific Crest Trail, another sign-saturated hiking highway. Still, when we stop at a small store for lunch supplies, Michael buys a topographic map. Mid-morning, we sit on the crumbling couch-sized trunk of a ponderosa to eat Fig Newtons and to read, in the same moment, inked contour lines and the growth rings of cut trees. We match the world to paper: that hill must be these concentric rings, that valley must hold this blue line.

I read Columbus's diaries and see that he sailed on the imagination of maps, then drew his own. He named his world, too, learning the Arawak words for the Caribbean islands and rechristening them for Spain. Maps and names framed the slow journey of Lewis and Clark, who left home with trappers' directions and drawings, collected impromptu maps from the people they encountered, followed native guides over old trails, learned and gave names, drew and redrew their own evolving visions of the continent. In one wilderness writer after another, I find an enchantment with names and maps tied fast to the love of wild places.

We know that we've come late to this world, that every place we go has long held human signs. And so we seek in those signs traces of other encounters with the land, of older, more intimate knowledge than ours. We live in a culture of paper and words, and it is our deep habit to understand what lies around us through their agency.

How much, I sometimes worry, is this habit a screen for possessiveness, a fancy velvet hat to cover a rat's snarl of greed and grasping? After all, Columbus's renaming of Guanahani as San Salvador, Bohío as Española, were acts of conquest followed quickly by genocide, and the journey of Lewis and Clark charted the path of Manifest Destiny. Surveyors and armies cross the land together. When John and I spent a year in China, we found that good Chinese maps were kept hidden from foreigners to maintain security—and, we heard, the best maps of all were those made by our CIA. Is it worse than inconsistent to love wild places and still think about them in ways so tainted?

Yet other stories are more reassuring. Hunters in the Far North can draw from memory detailed and dazzlingly accurate maps of their territory. The Pueblo, Navajo, and Apache people of the Southwest map their land and history with the stories held in place names. Australia's native people use names and songs to find their way over long and difficult routes, to remember complex and subtle landmarks. Everywhere in the world, it seems, we name and we map, we find, record, remember our way through our world with words and pictures. Perhaps whoever marked our trail to Surprise Pond was simply playing a mapping game—miming with outsized and funny gestures the kind of path writing we always do.

I browse through a bookstore in a mountain town and come across a file of glossy new waterproof topographic maps of Colorado. For no reason but nostalgia, I decide to look at the one covering the land where my camp stood. I check the copyright, ten years old, and the date of the latest revision—this year. The symbols of new maps leap out in black ink: tiny mountain bikes and cross-country skiers, stylized columbines marking scenic byways. Will the ski area appear, with its swarm of condos and giant hotels? All the new housing developments? The golf course and mansions that replaced camp's cabins and covered wagons? Expecting the icy water of reality, I check.

I can't believe it. This map has made a Shangri-La of my mountain valley. No change, no passage of time, no loss. Fifteen years after its disappearance, there is camp, the letters of its name, two tiny crescents of cabins and wagons, the ranch house and barn and tack room, microscopic memory dots. There's no golf course, no ski area, no building less than twenty, maybe thirty, years old. How could this be? I imagine another former camper at work in the office of the mapmakers, leaning over a computer or giant sheets of paper and remembering, plotting subversion. I buy the map to tape on my study wall. It makes me absurdly happy, this bit of illusion.

Maybe our wilderness maps are not after all so strange. Maybe, at least sometimes, they encode not alienation or grasping but respect, imagination, memory, love. Maybe when I sit in my kitchen or on a rock beside a trail, fascinated, enchanted, gazing into a sheet of that magical paper, crackling and slick with waterproofing or frayed and faded with many unfoldings, when I trace with my eyes and my fingertips the lines of roads and trails, wilderness boundaries and timberline, streams and cliff faces, when I savor the names of places as if they carried the taste of the earth, familiar flavors and exotic, sweet and savory, sharp spices to wake me up, Hummingbird Basin, Lupine Mountain, Frigid Air Pass—when in these moments I call into my mind a landscape remembered, present, imagined, when I place myself within that landscape, maybe then I am engaging, in the way of my own culture, in an ancient and universal activity.

I think of the topography of my hand, my fingerprints, the hills and valleys of my body, the twisting crevasses inside my skull, the bright tangled forest of neurons and pathways. Reading, walking, dreaming, on a trail, on unmarked ground, I follow the landwriting and find my way.

Pristine

Picture a field of new snow, soft and glittery like a smooth cloud of ground diamonds, clean and untracked. Its purity allows us to imagine that the earth below is equally unsullied, unspoiled. A virgin earth, an original earth, a primeval earth. *Pristine.*

Now listen for these words in conversations about wild places, and look for them when you read. You will find them everywhere. They are important words, words that have little to do with reality. They are the vocabulary of desire.

Let's imagine a perfect day in a pristine landscape—an unnamed desert canyon, a small dip among large mountains, an obscure lake in the Far North. We'll stay in the Americas, and we'll go alone. We'll leave roads and trails behind. We'll see no blazed trees, no wisps of bright plastic, no rock cairns, no boot prints. No fences or stray strands of wire, no circles of campfire-blackened rocks. No signs of cattle, horses, or domestic sheep—no tracks, no droppings crumbling into powder, none of the exotic plants they carry with them wherever they go to deposit in the ground they disturb. No smooth tree stumps with rings we can count. No tumbles of rusty rock rubble below holes picked in hillsides. In the waters, the fish will all be native, unstocked, unscarred. The sky will be absolutely clear. No planes will pass, and we'll hear no distant motors. Even with binoculars from the ridgeline, we'll see no signs of human life. We'll think maybe we're the first people to admire this view, to walk on this ground.

So many *nos* in this vocabulary! And so many *uns*! Uncorrupted, uncultivated, undeveloped, unnamed, unexplored, unfettered, unmapped, untrammeled! This landscape is a dream of fullness written in a language of emptiness, the yearning for presence spoken in words of absence. It is *pure: purus* for *clean*. It is *original: origo* and *oriri* for *source* and *to rise*. It is *primeval: primus* and *aevum* for *first* and *age*. In Latin, *pristinus* means previous, early, primitive, original. The words circle and enclose each other in a perfect sphere of emptiness and plenitude.

And illusion. Consider the sheer unlikeliness of our imagined day. No trails or human tracks, anywhere in sight? Maybe, if we travel far enough and in the right places, but not for long—and only if we ignore our own footprints. No signs of cattle, sheep, or horses? Many official wilderness areas welcome them. How long will our sky stay empty of smog, of airplanes and their ephemeral white traces? No satellites coasting across the night sky like unleashed stars? If we see a distant herd of elk, how will we know they haven't been tagged for someone's wildlife management study? How likely are we to encounter a grizzly, a pack of wolves, a herd of buffalo? A meadow of only native plants? Possible, but unlikely. If it storms in the afternoon, will the rain be acid? Will the ozone above us be pristinely thick? And if we know of some threat to this place—a proposed copper mine or the flooding from a massive dam—how will that knowledge intrude? The footprints of foreboding: how deep are they?

But these are the easier questions, the mere rim of illusion, the top layer of sediment in the Grand Canyon walls. Suppose we could turn back the clock. How far to the Age of the Pristine? Just before the Industrial Revolution? No. Before Columbus? Do we imagine an empty "New World"? But these continents were farmed and irrigated, their animals were hunted, they held large cities built of cut trees and quarried rock. In places they were overhunted, deforested, farmed to exhaustion. All the early European exploration narratives talk about encountering humans and their traces everywhere in the Americas. To imagine that in 1490 this land was untouched, we have to erase the history of millions of people and hundreds of cultures.

Must we go back, then, to a year before the first human step on this land? We've been here a long time—fifteen, thirty, maybe even fifty thousand years. Anthropologists keep speculating about earlier and earlier dates. When we first arrived, the land was very different. It was an Ice Age ago. There were saber-toothed tigers and giant sloths. What wildflowers did the wooly mammoth eat? Had the camels left for Asia? Were the knee-high native horses already extinct? And further back, when the dinosaurs lived here, what did the land look like then? The Ancestral Rockies had eroded, the new Rockies had not yet formed. What shape were the continents? I search through geology books. I think about the amazing exhibits at the dinosaur museum in Vernal, pictures of drifting continents matched with charts of evolving animals and chunks of fossils. The state of Utah floats around the globe like a leaf turning on a slow eddy. What does it mean to say dinosaurs lived *here?*

The dream of the pristine is a sweet onion of illusions, every layer of skin not quite transparent, every layer peeled leaving another below, the whole center always just a membrane away. Why, then, is it so powerful? What draws us so insistently to the vision of a world without ourselves, an earth innocent of our trespasses?

Sometimes, I know, we exploit the image of the pristine, its emotional allure, in the service of money—in advertisements for four-wheel-drive cars that will go right over the grasses and rocks to the edge of the pristine canyon; for new luxury resorts set among pristine mountains or rain forests; even, in a bizarre perversion of the troubling "virgin land" metaphor, for "feminine hygiene" perfumes. More often we evoke it in heart-felt defense of threatened places: write your senator *today* to protest oil exploration in this pristine valley. A proposed new ski area near my home, I read, would ruin a pristine forest—no matter that this land has been mined and lumbered and grazed for a century. Sometimes the language of the pristine allows us to ignore complexity.

We've done so much damage here—massacred the wildlife, decimated the seas and rivers, poisoned the waters, flattened the forests and prairies, wasted the topsoil, dirtied the air, punched holes in the atmosphere. I remember a television show about the Sea of Cortez, how I sat in open-mouthed amazement and delight watching a kaleidoscope of underwater wonders, a swirl of abundance and beauty I hadn't known existed, only to be stunned and nauseated in the last five minutes by the devastation wrought by bottom-dragging shrimp nets. Where, in such a world, is there room for hope?

Everywhere around us the wild garden recedes, leaving behind it a vacuum. We're pulled in its wake by what we've lost, by the elemental force of desire. So we dream of Eden, an original landscape straight from God, ours before our sin, our fall, our loss. This is at once an old dream for European cultures and a newly reborn one. Sometimes it's my dream, too, and I think it's not always a gesture of denial. Sometimes it's a vision of healing, a way to imagine we might erase our mistakes, a leap of faith.

Trudging

I plant my ponderosa stick in front of me, a short step down the steep, primitive Tanner Trail into the Grand Canyon, push it into the dust, and lean into it. I pull my right foot off the ground, just high enough to clear the rocks, swing it slowly toward the stick, wince as I feel the fierce pinch of kneecap against cartilage. Then the left. I try to keep my legs straight, shift my balance at my hips, hold my weight in my hands. For an instant I stop to rest. I lean against the cool red cliff on my left and look out and down into immense and silent space, breathe deeply, gather my strength. Then I start again. Plant, lean, right leg, left leg, rest.

I do this for hours and hours, over nine miles of trail that drop almost five thousand feet. Clock time gives way to body time, sun time, earth time. Under the monotonous brightness of noon, my shadow contracts, my skin bakes dry, the canyon flattens. Toward evening, the canyon deepens, my shadow grows long, I feel whispers of cooler air against my face. Sometimes, when the path levels out to cross a short plateau on the cusp of two eras, I can walk faster, swing along at a pretty good pace. Then when it narrows and drops over another cliff, another twenty million years, pain recalls me to nerves and bones, re-embodies me. I plant my ponderosa stick, lean into the ground, move one slow leg and then the other. I feel like an earthworm, measuring the trail with my body.

I'm thrilled to be here, I know it at every moment, but this journey is definitely a trudge.

Not that I'm surprised. In my family, trudging and sore knees are nearly synonymous. Mine started hurting when I was nineteen, at a party during my first week at college, when I danced so hard I needed crutches for a week. The next summer I climbed a mountain near Boulder with my boyfriend. Going up was no trouble. On top we held the camera out in front of us for a self-portrait of youth and innocence—two heads of long, straight, dark-blonde hair, two happy grins. On the way down, pain set in, knees went on strike. There was no trail, and once we hit the trees, deadwood lay everywhere. Every few steps I had to lean down, pick my leg up, and lift it over a fallen log. By the time we got home, both knees were swollen double. A series of doctors was baffled—arthritis, they guessed; take aspirin, they suggested—and I adjusted my sense of my limits. I *knew* this walk into the canyon would be painful and slow.

A few years later, I will find myself by my mother's hospital bed watching her recover from a knee-replacement operation. Her face will be as pale as old snow. No cartilage left at all, her surgeon will say—never seen one so far gone. She'll be strapped into a hydraulic contraption that bends her new knee just as far as she can stand it, a smidgeon farther, then straightens it out, every day a few degrees more, always right at the edge of tolerance. For the first day or so, she will have a button to press for an instant dose of intravenous painkiller. A couple of months after her operation, the next time my knees hurt, I'll see an orthopedic surgeon—twenty years have made quite a difference in sports medicine—and start physical therapy. A bribe to the genetic fates, a trudge insurance policy.

Of course even flawless knees won't stave off trudging. The other possibilities are infinite—as all wilderness travelers discover. Through deep snow, powdery, wet, ice-crusted, not quite stiff enough to hold a body's weight. Through swamps and marshes. Over ankle-turning tundra tussocks. In icy water, wild water. On high slippery scree, over boulders, across plains, hills, ravines, ridges, valleys. Along washes, arroyos, rocky creek beds, ambushed by thorns and cactus spines. Through stiff brush and trees tossed in piles of pick-up-stick chaos.

On foot, in small boats, on horseback. Carrying heavy packs, or light ones, or nothing. Shoulders aching, back muscles pulling in sharp spasms. Wrenched ankles, battered shins, palms and knuckles scraped raw. Blisters at many stages: seedlings, full ripe fruit, dried pods. Sunburned skin, hot to the touch and smelling a bit of scorched flesh. Split lips and the tingling, numbness, and daggers of frostbite. Swarmed by mosquitoes, black flies, invisible biting gnats. Invaded by ticks, leeches, viruses, and bacteria. Face into high winds, drizzle, hard rain, lightning. Snow, sleet, hail, blinding sun, darkest dark.

Sooner or later, nearly every serious wilderness journey turns into a trudge. Discomfort mixes with boredom and you think you'd rather be somewhere else, but you just keep going.

Sometimes I console myself with the thought that at least I'm in good historical and literary company, since trudging passages appear in nearly every wilderness narrative, in variations that mirror the trudgers' personalities. Consider this description of dogsledding in Alaska's Brooks Range: "It sometimes happens that three miles is thirty! Our trail led us down steep banks into sloughs, up steep banks out of them, across stretches of gravel bar, through forests growing on sand bars, with never enough snow to cover sand and gravel, the hardest pulling known to the dog musher. . . . I wondered how many more banks and sloughs and creeks and gravel bars there could be. I ached all over; I had a strong impulse to just flop down on one of those dry grass patches and stay there—I couldn't even feel what my feet were doing." Margaret Murie's directness and simplicity contrast wonderfully with the excruciating extravagance of Bob Marshall's account of backpacking in the same landscape:

> It is hard to describe how slow and plodding it really seems when you are out of practice and there is no trail and you have a 55-pound pack tugging on your headstrap and shoulders. You have hardly gone five minutes when the muscles in your neck are so sore that you know every step for the next six or seven hours will be pain. You throw off the headstrap to rest the neck and the pack pulls so violently on your shoulders you imagine it is turning them inside out. You go back to the headstrap again, pushing against it for all you are worth, perspiring freely in spite of the hour of evening, swatting at fifty mosquitoes which have lighted on your forehead and your cheeks and your neck, letting down your black mosquito net which instantly makes the whole world dark, pulling it up again when you almost stifle in the sultry evening, noticing suddenly that your ankle is sore where the boot has rubbed off the skin, stumbling over sedge tussocks, forcing your way through thick willow brush, sliding along on uncertain side hills ankle deep with sphagnum moss, neck aching, shoulders aching, ankle aching, on, on, on.

Remembering these passages, I wonder why I still want to go to those mountains.

Of course there is one pleasure in such episodes. They end. It's like the old joke: I love to hit my head against the wall—it feels so good when I stop. Perverse, yes, but all trudgers know the intensity of the relief.

There's also the pleasure of stories. Out of the longest and most insufferable trudges may come the most vivid descriptions. Thin and predictable in plot, perhaps, but with a distinctive style. Short sentences move step by step. Long sentences seem appropriately endless. Repetition is everywhere: we go on, on, on; I ache, ache, ache. The sled overturns on the ice three times, the ankle twists every few steps, false summit follows false summit. There's exaggeration: one mile of this seemed like five, my boots pooled in blood, I couldn't feel my feet. There's an appeal to the listener's imagination and doubt that it will be adequate: You cannot imagine what it was like. One trudge story calls for another, then another, each one mixing nightmare and farce, endurance and tedium, realism and hyperbole.

But I wonder if maybe there are deeper rewards from this experience, too, something more to account for our willingness to remember and repeat the real discomfort of wilderness journeys. I think about how my friend Brock puts himself in the way of such pain: "death marches," he calls his favorite hikes, looking delighted. Perhaps trudging is a kind of trial—of physical and mental strength, the mythic hero moving through the maze. If you persevere, you will reach your goal: the allure of a pure challenge.

And then there's the effect of sheer repetition. Time and rhythm hypnotize us into a different state of body and mind, blank and open. We may try to forget where we are, distract ourselves with daydreams, but when the trudge takes over, everything disappears except bodies. Follow your breath, the Zen masters say. Follow your footsteps. The trudge holds us firmly in the present, in the immediate material world. The earth grasps us with all its strength, invisible waves and particles of gravity, gravitons and gluons, clutching our feet and pulling us in, gluing us down. Trudging scrapes off our edges of concept and comfort, sands us down the way silty water scours river rocks, shapes us to *this* instant, *this* place.

Easing myself painfully into the Grand Canyon, I'm living its immensity. With the compacting of cartilage between my own bones, I *feel* the thickness of the strata, every step an eon, and with every brush of my hand against rock I touch the planet's past, its rough and vivid present. I'll *remember* this. Excruciating, insistent, intractable, my quintessential trudge. Gradually, the rhythm of my steps transforms the pain into a kind of ritual offering—to the dirt and rocks, the turquoise-throated lizards and the sweet cascading song of canyon wrens—and I don't have to try so hard to pay attention to what's around me. My walk becomes a kind of pilgrimage, a journey of my body deep into the earth's body.

I think about the snowy spring morning I watched a worshipper approaching a Buddhist monastery on the Tibetan plateau in China. She was ancient and tiny. She lay flat on the bare and frozen ground, then stood, stiffly, where her head had been, then lay flat again. Inching her way, measuring the distance with bones, cartilage, nerves, and skin, she must have hurt, too. I imagine she was filled with joy.

Grandeur

"What do you suppose the conquistadors saw when they looked into the Grand Canyon?" I ask John. "I don't know," he says. "Maybe just a barrier to conquest? A terrifying wasteland?" We start trading challenges: How about a tenth-century Anasazi potter? A good source of clay, or a well-protected place for a cliff house? What about a road builder from central China? A sheepherder from the Mongolian steppes? An Amazonian shaman? A small business owner in Zaire? It's fun to speculate, but it's clear to us both that we really have no idea. History and culture shape and limit everybody's vision.

We're driving toward the canyon ourselves on a short January afternoon. It will be our third visit, our first in winter, a few years after our trudge down the Tanner Trail. Driven by an attack of aesthetic greed, we're hoping for fresh snow and rushing to reach the rim by sunset. We know perfectly well what *we'll* see: GRANDEUR.

I'm having fun with my new knowledge. I've been reading about the history of European and American landscape aesthetics, tracing the amazingly straight line from the late eighteenth century to today. I've told John all about it, probably more than he wants to know, and now, as we look idly out the car windows at the passing pines and sagebrush, we're playing with words and images, testing our own cultural training.

"Let's make up the text for a brochure," I propose, "and see how many clichés we can fit in. We could call it 'Winter Sunset over the Grand Canyon.'" John's good at things like this—for a while he wrote travel articles for the mass market—and we both know the required ingredients. No need for paper, of course: this is a voice brochure. It goes something like this: "As the afternoon sun slides toward the far horizon, the canyon's vast and awesome spaces pool with delicate pastels. The majestic rock walls deepen and glow, molten and somber. Snow lies in small, soft pillows atop the long-needled branches of ponderosas, on narrow cliff edges where rock layers lie stacked like Navajo rugs, on the spires and crags that tower above the gloomy depths. Just before dark, the sky swells with brilliant tints of garnet, topaz, and opal. A Grand Canyon indeed!" We know what pictures we'll need, too: a Turneresque sky; the many-layered, snow-dusted canyon framed on at least one side by a frosty, gnarled, Christmas-card pine; maybe an old sepia-toned photo or a Romantic painting of cliffs and shadows. Some old-fashioned bits here, for sure, but nothing we haven't seen a hundred times in recent brochures, postcards, magazines, TV shows, car ads, pretty calendars, coffee-table books, signs at scenic overlooks.

Exactly how old, I wonder, are these images? How many decades would we have to go back before we'd be out of their necessary cultural framework? Would the history of canyon aesthetics match that of high, rugged mountains? It seems likely.

Until late in the eighteenth century, I know, big mountains were thought ugly and frightening, unfinished parts of the planet. I once read a vivid but apocryphal story in which Alexander Pope, traveling in a carriage through the Alps, had the shades drawn to shut out the horrible view. When such landscapes became acceptable to Europeans (and thus, later, to many Americans), they were not considered *beautiful*, a word reserved for gentle, humanized places—fertile river valleys, groves of fruit trees, small waterfalls framed by delicate ferns and orchids. They were *sublime*. Frightful, mysterious, awful, overpowering, savage, stupendous, desolate, dreary, inhuman: this vocabulary marked the way sublime landscapes evoked fear and amazement, wonder at the magnificent power of God, humility at human insignificance. Edmund Burke, the English philosopher who made these categories popular, said beautiful landscapes were founded on pleasure, sublime ones on pain. As Thoreau remarked about some waterfalls in Maine, "There was really danger of their losing their sublimity in losing their power to harm us." Vast chasms and white-water torrents, looming glaciers, sheer cliffs, each carefully framed into a picture with a blasted tree in the foreground: such places—such *views*—were sublime.

Places, in short, like the Grand Canyon. Its first official explorer and surveyor, John Wesley Powell, packed his narratives with the language of sublimity. So have many later visitors—including John and me. "Grandeur" isn't a word I normally use, though I often seek out such overwhelming landscapes, but here it seems right.

We show our annual pass at the park gate and drive the last few miles, enjoying the suspense of the approach—these forests don't even hint at what's ahead. How far would this one word stretch? Would we use "grandeur" to describe the Sonoran Desert, with its imposing saguaros? Maybe, though deserts, like jungles and swamps, entered the ranks of acceptable scenery too long after Burke to be included automatically. How about the view from the top of one of the tall mountains near my old camp, where high rocky peaks stretch to the horizon in all directions? Definitely. The Black Hills, vast, rolling, and buffalo-dotted, or Virginia's Blue Ridge? John, who grew up on Long Island, says yes; thinking of the Rockies and the Alps, I say no: too rounded, too green. The enormous flocks of sandhill cranes (12,000 of them!) and snow geese (24,000!) we've just seen wintering in New Mexico? A tough one. The word implies great size and scale, but it also suggests inaccessibility, a forbidding distance between the viewer and the view. It's a word about looking *at* a landscape that you would hesitate to enter.

We're here, and just in time for sunset. Exhilarated, we recite together the opening lines of Hopkins's sonnet "God's Grandeur," one of the few poems we both know by heart:

The world is charged with the grandeur of God.
  It will flame out, like shining from shook foil;
  It gathers to a greatness, like the ooze of oil
Crushed. . . .

A sign saying "Scenic Overlook," a parking lot, the end of the trees. We jump out of the car and head for the edge.

Disappointment. The clouds are thick and the light lackluster. Under the gray sky the rock walls look dull, and somehow the spaces I know are huge seem flat and uninteresting. There's hardly even any snow, just enough ice on the paths to make us walk very carefully and think about what would happen if we slipped too close to the edge. We see no brilliant light or somber shadows, no glowing colors, certainly no shining like shook foil. None of this should matter—the canyon is no smaller, no more accessible, no less astonishing or complex—but somehow, it seems, *grandeur* requires the right weather, the kind of light we know from paintings and photographs.

What's worse, the canyon is filled with haze—not with the pastels or soft blues I've seen so often in photos, but with a dull mist just dense enough to make me check whether my glasses are dirty. This effect doesn't bother John, who calls it fog and thinks it looks softly dramatic. Nor do the other tourists seem to mind: they lean happily on the iron railings taking pictures of the view and each other. But to my eyes the haze has the greasy sheen of smog—a possibility all too likely. A dirty view is not sublime. Smog and grandeur do not mix. Hopkins's poem, I remember reluctantly, follows that exuberant opening with these depressing lines:

. . . Why do men then now not reck his rod?
Generations have trod, have trod, have trod;
  And all is seared with trade; bleared, smeared with toil;

And wears man's smudge and shares man's smell: the soil
Is bare now, nor can foot feel, being shod.

I can barely make myself look into the canyon. I feel betrayed—not just by my own expectations, but by this unwelcome reminder of human interference even with the most magnificent and inhuman of landscapes.

We eat a dull meal in an echoing cafeteria, waste a half-hour wandering aimlessly around on dark roads, and go to sleep early.

Overnight a storm blows through. It brings a soft, glowing early morning light, clean cold air, and a layer of fresh snow for every mesa and pinnacle below. John has disappeared by the time I wake, so I walk alone to the rim and start slowly and carefully down the Bright Angel Trail. It's wide and heavily used, the main highway down to the river, but this morning my footprints are the first in the new snow. It's very quiet. A trio of ravens flies past just beneath me and I can hear the whoosh of wind in their wings. The canyon is restored, ancient and newborn.

Hopkins's poem comes once more into my head. This time it's the closing lines, lines that have always seemed to me to echo the lost past, an older and much stronger faith in rebirth than I can usually muster. I say them softly aloud to the squawking ravens:

And for all this, nature is never spent;
There lives the dearest freshness deep down things;
And though the last lights off the black West went
Oh, morning, at the brown brink eastward, springs—
Because the Holy Ghost over the bent
World broods with warm breast and with ah! bright wings.

Why "*dearest* freshness?" I'd always wondered. This morning I think I know. Maybe it's just the soft light and the new snow. Or maybe it's the reward of my painful walk down to the river and back up a few years ago—on a trail so fine and broken we could hardly find it from the rim after we came back out. That slow journey, those long hot dreamy days by the icy river, the delicate melodies of canyon wrens and quick turquoise lizards, the tracks left by beavers dragging their broad tails—these things seem to have changed my relationship with this place. Certainly I'd hesitate before I'd take that trip again, but it makes a surprisingly big difference to have been *in* the canyon. By walking into the sublime, I seem to have broken the frame of the picturesque, transformed one kind of landscape of desire into another.

I watch the bright ravens dance over space. I imagine the way snow must lie on the tamarisk branches by the river and wonder what animals are brooding below me. Though I can hardly believe it myself—I doubt I've earned it—the word I'd use for this place on this morning is not "grandeur" but "intimacy."

Misery

Beyond trudging lies misery.

I can't see a thing. Only my feet, about a yard of trail, and the backs of John's legs. My glasses are foggy, my cap pulled low. It's raining. I'm sopped and chilled, sweating and shivering. Several inches of mud and standing water cover flat pieces of trail; steep pieces have turned into streams. Rocks lie everywhere, sharp ankle-ambushing rocks, rounded ones so thick with moss that ferns, even trees, sprout from them. All of it is crisscrossed with six-inch roots and hip-high fallen trees. Mushrooms smother every surface. Everything is wet, everything is slick.

Over my head is the back half of an eighty-pound canoe. Forty pounds of it rest on the top bone in my spine, the unpadded one that sticks out just enough to make a shelf. I've added cushioning—a balled-up fleece jacket—which helps when it isn't sliding sideways or down. My arms are angled sharply up, elbow joints pinched and awkward, and I'm grabbing on to the thwart, but I haven't got the arm muscle to hold any of the weight, only enough to balance it. The muscle behind my right shoulder that cramps up all the time at home is one spasm of pain. My left arm tingles as though it has fallen asleep. Sometimes I try to hunch my shoulders enough to take the weight off my spine, but then the cramp sharpens. When John takes a long step, the thwart hits the back of my neck and yanks me forward. When he swings to the left, I'm swung to the right. When he stops, the thwart slides off its cushion and down my spine, and I feel the weight in the small of my back. If I don't take steps exactly the same length as his, I lose my balance. It's a nightmare of dancing, a nasty parody of a romantic vision of marriage. I walk as straight a line as I can, right through mud and water, up the sides of cracked boulders, down root ladders. I look only at our feet.

I'm covered with bruises and sore spots. Shoulders, upper arms, elbows, forearms. Wrists, palms, fingers. Hip joints. The tops of my calves and the outsides of my legs just below my knees. My toes, my soles. The wind kicks up and John says, "Better put your jacket on. Don't want those shoulders to tighten up."

The routine is relentless: Put the canoe in the water. Sling in two heavy backpacks and arrange them on the bottom under the thwarts, straps and frames down to keep the contents dry. Push off. Paddle five or ten minutes into the wind. Glance at the view: trees and rocks, all grayed with mist. (The sun hasn't appeared once in the three chilly late-September days we've been here. This is the best time of year to come, everyone says. In the summer it's hot and muggy and the bugs are awful.) Land. Haul out the backpacks and canoe, sometimes up steep boulders or over piles of sharp rocks. Hike fifteen minutes with backpacks, waterproof ammo box, paddles, pads. Hike back, sucking on peppermints, enjoying the temporary respite, trying to be cheerful. John jokes about when I'm going to burst into tears (I won't—I never cry when those natural tranquilizers might actually help) and wonders what we would do if our limited view from under the canoe suddenly included four tall, knobby moose legs. Arrange fleece jacket into neck cushion. Heave canoe over head. Trudge. Put canoe in water, load, paddle with cramped muscles, land. And we're doing this the easy way, not the purist's way. We have real backpacks with hip straps instead of shapeless canvas and leather sacks hanging off our shoulders. We're making two trips per portage instead of one.

We're in a hurry, too, trying, though it's clearly hopeless, to keep up with our friends Lisa and Larry, who have brought us here to the Boundary Waters between Minnesota and Canada. When we arrive at a portage, they're starting their second, canoe-bearing, trip. (They're carrying the heavier rented boat, leaving their own lighter one for us. We feel a little guilty about this, but mostly grateful.) When we reach a lake with our backpacks, they're disappearing into the mist, paddles rising and dipping in perfect unison. If they get too far ahead, they wait just until they see us coming, then take off. On one long portage, they come back to help us with our canoe, explaining that this will make us all equally tired. They're trying to be tactful. "We don't know many couples who like to do this kind of thing," Lisa says, sounding perplexed and a bit forlorn. "We don't mind slowing down a little for you guys." I set my jaw, shake out my right elbow, adjust the canoe on the top of my spine, and tell John I can go a little faster.

"I hate this," I think. "This is awful. This is torture. How could anyone like this? Why would anybody want to come here? Why did I want to come here?" Everything hurts. I don't like the cold rain and thick mist and dark clouds. I don't like tangled trees or piles of mushrooms or the monotony of small lakes, small hills, endless forests. I feel claustrophobic. I don't like to hurry, and I don't like not being able to see. Misery is blinding me, cutting me off. My own body is taking all my attention. The land is a blur. I feel profoundly off balance and out of place. I have no idea where I am.

Three days of this and finally I'm familiar enough with the routine to start thinking. I'll analyze how I feel, I say to myself. I'll isolate the ingredients and pin them down in my head. Maybe it will help.

The weather. Western sunshine addict returns to hated Midwest, disappears into monster rain cloud. It's not casual, my preference for sagebrush over ferns, sand over mud. John and I have avoided hiking and camping in soggy weather for years, but this time we're trapped by airplane tickets and complicated schedules. When the sun breaks through on our fourth day, my mood shoots up.

Then there are expectations. Mine had been high but ludicrously inaccurate. Slow drifting over calm sunny lakes, loons and moose, maybe wolves and the northern lights. The intrigue of a new landscape, the familiar comfort of a canoe. A little bit of portaging each day, just enough work to earn my supper, plenty of stops if it turned out to be hard. I'd asked John and Lisa how much carrying to expect, but their answers had been vague enough to leave this daydream intact. And I'd thought I'd be resting the canoe on my shoulders, not my backbone.

Ah, yes, shoulders and spines. Clearly I'm not in adequate condition for this kind of travel. After a summer in the Rockies, my legs and lungs are strong, but what I need here is arm and shoulder muscles. I should have spent weeks doing pushups and moving boxes of books. My bone structure isn't helping. When I get home after the trip, I'll try to show family and friends which vertebra the canoe sat on. Nobody else has one like mine—one that sticks out without natural padding.

The company's clearly a factor, too. I feel forty years older than Lisa and Larry, not twelve. One morning we watch them paddle crisply away from us and John says, "You know, Larry was captain of the triathlon team at the University of Michigan." Lisa was a top-level triathlete, too, it turns out. When the trip is over and we're all down to T-shirts, I'll suddenly see what triangular backs they have, what rounded muscular shoulders. This information makes me feel much better—in retrospect.

And then there's pace—maybe the main ingredient. I'm realizing just how important it is to me to set my own. All those years as a camp counselor trained me to move at a speed sustainable by a twelve-year-old girl. In town I'm a fast walker, but in wild places I'm just about the slowest traveler I know. I like to mosey and dawdle. I like to drift, take silly exploratory detours, look for birds and wild animals, stop to study pieces of bark and seedpods.

In a canoe I like to be pushed and spun by wind and water. Everything that lures me outdoors disappears when I have to rush.

Finally we start traveling separately between meal rendezvous and then take different routes for the last day and night. As soon as John and I slow down, my trip changes. A moose appears for a split second, huge dark antlers and massive body, then crashes away. Pairs of loons float close to our canoe, call softly to each other. Beavers swim across our path. Through the early morning mist we hear what must be wolves. Though by now we're retracing our route, heading out, to me everything looks absolutely different. It's not just the new sunshine. It's the quality of my attention.

Now when my shoulder cramps hard and the boat slams into my spine and shoves me off a wet root into a frigid puddle, I distract myself by thinking of the stories I can tell. Misery always makes good stories. I should get a lot of mileage out of this trip.

And I do. My family and friends laugh when John and I act out the canoe routine. They exclaim and cringe and groan when I describe the varieties of pain and discomfort. "I don't think I'd like that," says my athletic, mountain-climbing niece Léa, sounding thoughtful. "Oh, SueEllen, that sounds *awful!* I'd have cried the whole time," my friend Nina commiserates. "You mean you slept *outside?* On the *ground?*" asks John's mother, a city person. They're happy that they didn't go with us, and we're heady with relief that the trip is finished. We're all revelling in what amounts to a benign, reassuring tale of survival. We were never in any particular danger and we've done no lasting damage to ourselves; in fact, by the time we get home, we're barely even stiff. Misery fades and the fun of stories take over.

But I start to notice something else, too. Not everyone has the same reaction: a few listeners look slightly startled, then slightly amused, then tactfully entertained. Clearly the trip doesn't sound all that miserable to them. I remember once hearing a friend describe in outraged hyperbole his horror at discovering that when you backpack in the mountains you sometimes have to cross fast, icy streams on elevated logs. I'd done this maneuver often enough to react less with empathy than with a kind of Olympian amusement. When I see this same response in a few of my friends, I check their shoulders: sure enough, rounded and muscular.

One of them, Don, is a western mountain climber who recently moved to Minnesota and started canoeing in the Boundary Waters. His canoe, I learn, only weighs forty pounds—but then he carries it alone, and he uses one of those big shapeless bags instead of a backpack: a purist. I figure maybe he can explain to me—in terms I can understand—the appeal of this kind of travel in this kind of landscape. "So what's there to *like?*" I ask. "It all seemed so monotonous, so claustrophobic." "One thing I like," he says, "is how hard it is. It's shown me that mountains aren't really all that difficult. Also I like the challenge of the monotony. It's harder to read the landscape, harder to find your way. You have to look really carefully for details and differences. It teaches you how to see subtleties." This makes sense to me: when I stopped being too miserable to look around, I'd started to notice the same effect.

Another one, Mark, I meet for dinner before he's to give a talk and slide show on his recent expedition deep into an unmapped part of the Tibetan Himalayas. I know better than to tell this major-league adventurer my mild canoeing story; I only mention that I found the trip really hard, even miserable. He laughs and says, "But misery is the reason to go! It raises your consciousness." "Not mine," I blurt. "I was too uncomfortable to be conscious of anything." He laughs again and pats my back: "You're sweet."

Misery, we decide, covers a broad scale. I remember how an old boyfriend used to say that suffering is worship. That never made sense to me; it still doesn't. I guess I don't have the temperament to be a serious worshipper, a true pilgrim. (I want to add: a martyr, a masochist.) Mark and I talk about the Tibetan pilgrims we've seen approaching monasteries. That old woman I watched in western China: How many days had she spent prostrate on the frozen earth, inching her way toward this holy place? How far had she come, through what late-winter storms? Was her journey one of thanks, supplication, penance, praise? She must have been cold and exhausted, hungry and sore. Was she really full of joy? Was she miserable? What did she see?

Paying Attention to Detail

For years, it seems to me now, I passed my time in wild places in an unfocused blur of pleasure. I'd sit and gaze, mesmerized, at the swirling prisms of a waterfall, the rush and melded colors of a campfire, the endless rough texture of the world seen from a mountaintop. Strong summer scents would fill my throat—hot dust and horses, sagebrush and cattle in rain, sweet wild roses and clover and the pungency of blue spruce. I'd talk, sing, daydream, think nothing at all, breathe in time with my footsteps, feel the warm sun on my skin or the cold sharpness of hail, float through a space of sensory luxuries. Everything was intense and wonderful, all the effects somehow large.

Slowly, summer to summer, the swirl around me settled and came into focus, each clear circle anchored by detail. Early on I learned to spot wild strawberries, wiry red runners and round saw-edged leaves, hidden soft pebbles of concentrated flavor. Slowly these tiny zones of clarity gathered others. Wild raspberries, always on hot, steep slopes, scraping my hands and forearms as payment. The dark eyes and fine white powder of aspen bark. Rarely, a feather, orange flicker, sky-blue mountain bluebird, brown and tan hawk. Mostly, though, I learned to see individual flowers. Fairy trumpet or scarlet gilia, red flutes on tall, slender stalks. Alpine forget-me-not, minuscule and incandescent blue, cushioned like moss on the gravel of ridgelines. Mariposa lilies, monkshood, paintbrush. Each one distinct.

It was like finding my way into a new world, leaving behind in the nostalgia of remembered pleasures those impressionist landscapes of my teens and early twenties, my Wordsworthian delight in those soft, blurred days. A sharper world, more intricate and precise and vivid, abundant recompense. Paying attention to detail, I'm focused on what's right around me, tied to a precise place and moment, sometimes admitted for an instant to that paradoxical and much desired state, the pure present.

It's an easy and rewarding discipline, paying attention to detail. It coaxes me out of that peculiarly abstracted self-absorption that can grow from my body's discomfort or the clamor inside my head, from fretting about work or the future or the past. Plodding breathless up a familiar ridge in September wind much too cold for my bare ears and face, I concentrate on the small things around my feet: late asters, snow patches with inch-long tracks the shape of human feet, newly excavated holes, mysterious translucent seedpods, splashes of mica, scattered droppings from deer, elk, bighorn sheep, marmot. Short days lengthen when I watch the drifting shadows on cliff faces, when I sample the textures of bark, grass blades, pinecones, rock. I watch birds, not just the flashy ones I can name but all those little balls of brown and gray that hop among fallen leaves, and I try to sort out their songs and chirps from the murmur of wind, water, insects. Once I saw a hummingbird's tongue, a gleaming silver filament flashing out and back.

In a big landscape, detail centers me, holds me to the particular and my place in it. Vast views are balanced and clarified by the high shriek of a hawk, the amazing palette of lichens—chartreuse, orange, scarlet, tobacco, charcoal, sage, platinum. The desert may stretch a hundred miles on every side, but I am here, in this fragrant sandstone scoop, under these frosty juniper berries, where the wind and the grass have drawn circles in the warm sand and some small bird has left delicate tracks.

Detail makes small landscapes expand, too. Sometimes I lie flat on the ground, maybe take off my glasses to shrink my field of vision, and a scene will come into focus that I'd never have thought to see: on a smooth gray rock blanketed with ruffled gray lichen, a miniature gray spider grooms its fuzzy legs. Or I look through my binoculars at things not so far away, just distant enough for the lenses to handle, and space will fill with that pile of fallen leaves or that new cone forming on the branch of that fir. At twenty feet a blossom is a patch of color in a broad sketch of shape. Through the lenses, each is a painting by Georgia O'Keefe projected against the sky.

Once in a while this magic happens with only the lenses in my own eyes. I'll be looking idly about, on a hill, say, in trees, just above a small sunlit marsh. On my right will be a freshet, for the snow is just now melting, and on my left another, each about a foot across, flowing together a step or two downhill from my feet. I'll sit quietly and pay attention. On my right the water falls in soft swirls of ponds, easy ripples, big slow bubbles floating over still pebbles. On my left it's faster, white water, whirlpools, roaring waterfalls, Niagara, Victoria Falls, boulder-tumbled chaos. Insects live here, and amphibians (I can hear a frog somewhere), fish and mammals. Moss hangs everywhere like cool, dense forests of ferns; grass blades look like high bamboo, fields of fallen needles like the tree-strewn slopes around Mount St. Helens, flung flat and scorched brown, black, gray. It's all the water in the world, right here, and every confluence, rushing down the steep slopes of all the mountain ranges, cutting canyons and broad valleys, slipping silently through swamps, and then to the sea.

An angle of vision. How can we ever know what distinguishes our own? When my friend Susan asks me to think about leading some nature walks at the local environmental learning center, my instant reaction is, "But I hardly know anything about it. I'd have nothing interesting to say." Her reply astonishes me: "The way you *see* things is so unusual—at least it's very different from what most of the people who work out here seem to see." Yes, I want to answer, they know things and I don't. I'm such an amateur—curious but with a terrible memory for specifics.

What could she have meant? If there's anything certain about my own vision, it's that to me it's absolutely *usual*. Thinking about this riddle, I realize one thing at least: it's no accident that Susan spoke of the way I *see*. Though lately I've been trying to pay more attention to my other senses, still my clearest perceptions of the world are insistently visual—maybe because I have more words for the nuances of shape and color and light and can give them the stability of names, or maybe because I'm always compensating for my myopia. I'm always amazed when a hiking companion can identify a bird by its call—I've usually heard nothing to notice. Walking along a path through a meadow I'll be ambushed by the intensely familiar smell of something—but what?—and have to work to recognize it as clover. My skin, too, remembers, but in an uncharted realm beyond language.

When I think about what I've studied, it makes sense that my world should be filtered and shaped by words and eyes. Literature is my job, and so of course it frames what I see. Hopkins, Wordsworth, Wallace Stevens, Annie Dillard, Barry Lopez, Peter Matthiessen, Lewis and Clark, William Stafford—my list grows as I write it. In college I did a second major in art and art history—Albert Bierstadt and Ansel Adams, O'Keefe and Turner, the impressionists and photorealists. I sharpened my eye for shapes in a sculpture class, for composition and focus and light in photography. Not long ago I took a watercolor class. It was a week before I understood what my teacher meant by "values," and a month before I began to see them. And figuring which pigments would mix to make a particular color! I'd never have thought it could be so hard. My paints still look like mud, but my eyes are more sophisticated now. Light and color and shadows and shapes—all these basic building blocks changed completely when I tried to look around me as a watercolorist.

New landscapes shift vision, too, and the places I've lived and visited must have helped shape mine. Lush Houston and Virginia, dry windy Wyoming and cloudy Ohio, Thailand and Mongolia, Italy and Siberia, China's rock deserts and rain forests and cities so crowded you can't take two steps down the street without being jostled—in places so different from each other and from home, I've been sure I could *feel* my eyes adjusting, refocusing, learning to see again.

Sometimes I'll pick up a bit of natural history that will have the same effect. I read on a sign somewhere in Utah that the desert sand is coated with crusty charcoal-colored stuff that's alive and has the wonderful name of cryptogamic soil, for "secretly married"—and suddenly I see it everywhere around my feet. The spurs on blue columbine leap into focus when I hear that they collect nectar. I've always liked to look at those squiggles on trees that have lost their bark—tree hieroglyphics—but they're somehow transformed when a friend tells me that they're called "galleries" and that foresters and entomologists can tell by their shapes what kind of insect made them. Photos of what birds and insects might see amaze me and challenge me to cut my own vision loose from habit—prism vision, hundreds of repeated images, kaleidoscope worlds, color patterns so strange that I don't recognize familiar flowers.

John writes a children's book about volcanoes and I learn enough to start seeing them everywhere. The northwest corner of New Mexico—right on my route between home and college—sprouts a field of cinder cones. The cliffs around a friend's cabin stand newly revealed as eroded stacks of tuff and ash. A pass I'd driven over dozens of times suddenly has the shape of old volcanoes and lava plugs. Even one of the mountains I looked at every day at Skyland—and climbed more than once—turns out to be volcanic. Ten years ago I'd have guessed there had never been volcanoes in Colorado. Now I know better. A new eye for new details, and the planet is transformed.

Still, I'm always realizing again how little I see, how partial even my field of vision remains. I call my friend Bill, a wildlife biologist who has spent decades working with deer and elk, and ask him what he sees when he looks at a herd of elk. I expect a scientific-sounding answer, something about reproduction rates or carrying capacities or diseases or territoriality. "I don't see a herd," he answers. "I see a collection of individuals. Each one looks different to me—different faces, shapes, colors. They're as separate to me as a group of people would be to you." Old cows, he says, have long narrow faces. They're the great-great-grandmothers, the matriarchs, the ones who keep and share information with their daughters and granddaughters, the other members of the group. The colors are different on each one. "Sexy isn't exactly the right word," he says, "but some of them have fantastic eye shadows, really lovely."

Sometimes I feel sharp and alert, zoomed in on the details around me, present in the present moment. Other times I know this is an illusion. A piece or two at a time may be in sharp focus, but the rest is always a blur. Elk eye shadows! Amazing.

Wildflowers

I love them. Like a hummingbird living on nectar, I fly to wildflowers, immerse myself in their colors, perfumes, fabulous shapes. I say their names to myself during the cold brown and white winters and again when I see them each summer. In exotic places I gaze at strangers, and at home I greet old neighbors, kin.

My interest is emphatically amateur. Sometimes I use field guides, but never rigorously. I'll learn to use a key when I retire—or later. Why would I want to count stamens and pistils? And taking a flower apart makes me cringe. So do Latin names, most of them. A few are tolerable—*Achillea millefolium*, for instance, since I like the story of Achilles using this many-leaved plant to heal his wounded soldiers—but plain yarrow is also fine. Otherwise they just seem long, pretentious, downright ugly—and not necessarily all that precise, either. *Eritrichium elongatum* (or *nanum*, or *aretioides*, depending on which book I check) for lovely little alpine forget-me-not? Or the family Scrophulariaceae, for such beauties as Chinese houses, Indian paintbrush, butter and eggs, seep-spring monkeyflower, and all the penstemons? Not that the Figwort family is much more euphonious. Snapdragon seems much better.

I prefer the common names. They cover the scale of tones—

melodic, poetic, fanciful, evocative, silly, lumbering, plain. Sky pilot and death camas. Jacob's ladder and pearly everlasting. Mariposa lily and snowball saxifrage. Heartleaf arnica and flax. Someday maybe I'll see a fetid adder's tongue or a phantom orchid or a forktooth ookow. So often there's a choice, a word bonus. I like glacier lily more than dog-tooth violet: nothing doglike or toothlike there that I can see, but they do climb with the melting snow at about the same time I do. Depending on my mood, I choose parrots-beak or sickle-top lousewort. You can invent names, too. Peppermint flower, raspberry parfait and vanilla paintpots, grape-jelly penstemon, dirty sox.

In size wildflowers cover the scale as well—some as wide as a freckle and as short as moss, others well over my head and as big as my face. And their shapes are endless and astonishing. Simple ones can be very satisfying, a pause and deep breath for the eyes: waxy white sand lilies with their splayed pointed petals, sky-blue harebells, oxeye daisies, lavender asters. But complex shapes rivet me, defy me to do them justice with words. Columbines, of course, and glacier lilies. Shooting stars. Fringed gentian. Elegant and graceful, these flowers fill three dimensions and burst into a fourth. They're good evidence for the old theory that God must exist because the world is so intricately designed.

I'm most amused by the miniatures, intricate renditions of the world's larger furnishings, maybe because they evoke my childhood love of dolls and dollhouses, of fairy tales and stories of children riding on bumblebees, taking baths in thimbles, floating down rivers on leaves. Fairy trumpets and pussy toes and bright yellow monkeyflowers. The deep purple blossoms of monkshood. (One good thing about field guides is their occasional odd detail: a European version of this flower was wolfbane, a shield against werewolves. Their shape makes this seem logical.) Summer marshes fill with stacks of tiny fuchsia elephant heads, each with two floppy ears and a rounded forehead stretching into a long trunk, each trunk curled at the tip with a flourish. Imagine the possibilities: swaying layers of crocodile, armadillo, or polar bear blossoms, a raven flower with glossy black wings, a baseball cap lousewort . . .

And the colors! I drench myself in them. They're rich, glowing, light as much as pigment, waves more than particles. Their variety seems endless. Blues brighten through larkspur, fringed gentian, and lupine, to the light-washed tint of the palest columbine. Purples branch off and redden. Most yellows are clean and bright like lemon peel, though some veer sideways toward orange, then red. The mixed-up species of paintbrush alone cover a rainbow—cream to pale yellow, light to bright orange, scarlet to raspberry, always tipped and edged with greens. Reds start near black with king's crown and fade to the palest of veins in evening primrose.

Wildflowers stun me with their beauty all the time, and with their character. A solitary plant set off by itself to be admired, an intimate landscape designed by a wild and invisible gardener, great sweeps of colors. A spiny clump of cactus barrels covered in vermillion under the pale delicacy of desert cliffrose. Lupine, paintbrush, and sage beneath silvered juniper berries. Under aspens or on sunny hillsides, blue columbines billow like clouds, their stems and dainty leaves almost invisible from any distance, their big blossoms swaying in the slightest breeze. Late in the summer a solitary survivor squeezed between boulders high above timberline has the aura of the rare. Lovely sky pilot, up close, smells faintly of skunk. Blankets of sunflowers, all facing east, stain whole slopes bright yellow, and wet meadows shift between blue and purple all summer long. On a high summit, alpine forget-me-nots glow like tiny sapphires.

My favorite flower walk is in the place John and I call Hummingbird Basin. Sometimes when we go there we count kinds of flowers—just the ones we can name or keep straight in our heads, just what we see along the trail as we go. We've come close to seventy.

The trail starts in the shade of spruce and fir, where mountain bluebells droop over geraniums, heartleaf arnica, sky pilot. Then the trees give way to flower meadows, several ambling hours' worth, where the variety and abundance are stunning.

I walk in flowers up to my shoulders. Hummingbirds zip everywhere—among the larkspur, in monkshood where rivulets fall across my path on their way to the river. When I wear red they sip at me, too, at my cap or my T-shirt, even the frames of my sunglasses. Queen Anne's lace and cow parsnip float by at eye-level with the flashy plumes of skunk cabbage. Shorter flowers reach my knees or thighs. Harebell and parrots-beak, sunflowers of all kinds, flax and columbine. Paintbrush in a half-dozen colors. Death camas and asters and bistort. The scent of lupine is strong. I pause, breathe in, breathe again. Grape-jelly penstemon and a pale cousin. There's no ground visible, and very little grass, but leaves frame everything. Brookcress, elephant heads, king's crown and queen's crown, monkeyflower, the small white stylized blossom that looks like a Zia sun. Sometimes all the colors are mixed together like a pointillist landscape, one flower and one tint at a time, the whole an iridescent shimmer. Other times they're spilled in streaks and pools, ribbons of white, splashes of scarlet, midnights of blue.

The colors stretch to the red rock rim, to the brilliant sunlit sky. Near the ridge top the tallest blossoms are just inches off the ground, trimmed low by winter winds, hugging the shrinking remains of the snow. Buttercups and marsh marigolds, moss campion and alpine phlox, yellow and blue violets. When I reach the ridge and sit to look at the view, I glance down and find that my jeans and shirt are streaked with yellow, from my ankles to my neck. I'm covered in pollen, spilling over with nectar.

Wild Animals

At dusk, in an empty part of Wyoming, a bobcat lopes across the road in front of our car, there and gone in an instant. We scramble out, dash to the side, look down a brushy gully. One more flash: gold and cream and brown, muscular, graceful, fleeting.

John and I are driving east from New Harmony, Utah, straight into the morning sun. Ahead, yesterday's snow glows on the still shadowy cliffs of the Kolob, at the northern end of Zion National Park. Already today we've spotted a dozen bald eagles, so we're paying attention and our windows are open. The unmistakable song of a western meadowlark—the first we've heard since last summer—pulls us to a stop and draws us backward until we see the singer, yellow breast and black collar perched on a fence post. We drive slowly on.

"Oh, no," John groans. We stop and back up again. It's a doe, caught somehow in the fence along the road. At the sound of our car, she moves, jerking awkwardly in place.

We walk toward her as quietly as we can and wince when she jerks again. She's caught her right hind leg in the top wire, as high as my shoulder, and she sprawls head down in the dirt. She lies still. I grasp her leg and hold it steady as we try to feel the path of the wire around her ankle. Luckily, it's one of the few plain strands in this fence, not one of the barbed ones above and below it, but it's pressed so deep into her flesh my fingers can't trace the whole snare.

She stays quiet as we try to ease the wires, but they are too tightly woven, not enough give anywhere in the fence. Her leg is warm, a thin layer of skin over bone and tendon, arteries and nerves, about as thick as my own wrist.

Hummingbirds zip around me, flashing green and purple, copper and vermillion, darting close to check for nectar in my red T-shirt. At first I flinched when they buzzed past my face, but now I hold my breath to feel the breeze from their wings.

This morning Mary and I startled a peregrine falcon standing on a tall stump. It startled us, too, as it flapped ahead of us down the trail to a taller tree, then, as we followed, to another. Now we're watching a pair of them circle high against the cliffs, harried by other birds too small, from here, to name. Their scream is longer than a red-tailed hawk's, high and piercing. We're in low beach chairs, barefoot, settled with binoculars and glasses of wine between the cabin's back porch and a clump of willows by the creek.

Mary stops speaking midsentence, points to our left. Just emerged from the willows is a weasel, nut brown, dark pointed face, long tail, pale yellow belly—the essence of sleekness. He stands still and looks at us. In his mouth he carries a very small, dead mouse, soft and thick like a miniature buffalo robe. We freeze. A curve or two at a time he moves closer, right past Mary's chair. Three inches from her foot he drops the mouse, stands upright, speeds toward my bare feet. Pure reflex: I jump, squeak. He dashes to one side, we move quickly a few steps to the other. Eyes still on us, he leaps out to the mouse, picks it up, zips under the cabin, vanishes.

Amazing. Face to face with a weasel. A peregrine. Hummingbirds. Like subatomic particles we collide, bounce back, collide again, energies in a vortex, sharing the space.

One year later I sit in the same chair, in the same place, writing these pages, and a weasel appears, surely the same one, though now sans mouse. Magic. Again he runs toward my bare feet. This time I hold still until he's almost to me, wiggle my toes, and he's gone.

At the beginning of the nineteenth century, many millions of buffalo lived on the Great Plains. By its end, there were almost none. Some in a remote part of Alberta, a couple dozen in Yellowstone: so very few escaped being shot from train windows for sport, slaughtered for profit, systematically wiped out as part of the war against the Plains Indians. There are more again now, of course, but they're no longer truly wild.

The small herd in the valley below me, here in this corner of the Black Hills, is protected and managed (most of the fifty or so calves will be sold at auction later this summer), but lying here on this warm May afternoon I can begin to imagine.

Vast fields of ungrazed grasses would have been so tall in some places that if you were walking, you could not have seen over their tops. There might have been, on the far horizon, the thin blue smoke of a wildfire or of campfires. Where the grasses were grazed low and the ground trampled to dust by thousands of hooves, bare stretches would have been covered with prairie dog mounds, mined everywhere with their elaborate tunnels, space shared and visited by burrowing owls, snakes, black-footed ferrets, weasels, coyotes, wolves. Elk and grizzlies were Plains animals then too, browsing on high grasses, wild fruits, stray buffalo calves. If they had looked at the right moments from their train windows, my great and great-great grandparents, moving to Kansas and Colorado, could have seen such sights.

On a May day, I imagine, the scene before me right here might have been much the same as it is today. The grasses are newly and intensely green, the loudest sounds those of the songbirds. The buffalo, more than two hundred of them, lie or stand quietly, basking in the sun, the still tiny calves close to their mothers. Down the hill on our right, two coyotes step slowly, gleaming gray, matching the occasional sagebrush. They drop into the meadow, circle the edge of the herd, looking casual, relaxed, flowing like liquid through the grass. Once they move too close to a calf, and the mother surges to her feet and charges toward them, just a few steps. They veer off and walk nonchalantly on across the stream, and she settles back into rest.

The coyotes disappear into the trees. The buffalo remain, undisturbed. Illusion.

It's clear that we can't free the doe with our hands only, so John jogs back to the car to look for tools. I stay at the fence, hoping to keep her from pulling the snare tighter.

I rest my wrist carefully between barbs and try to hold my hand steady around her leg. When I shift my weight, her skin slips a little over the bone, so warm, so taut, so thin to bear her weight, jump over fences, spring across meadows. I talk to her softly, surprised and relieved that my voice and touch don't seem to alarm her. She stays still. My hands are shaking.

Twice, sudden noises set her scrambling—a car passing by, then its door slamming shut as the driver stops to help. She can't really move, but her three free legs flail in the air, scrape at the dirt, as she tries to lift her head and breast off the ground. Each time, I grasp more firmly and keep talking while she quiets.

As I wait, my vision narrows to my wrist, hand, and fingers, wrapped around her ankle. The vast field of sage, rabbitbrush, and grass, the mountains and vivid sky, the lovely call of meadowlarks—all this fades out of focus.

I glance only once, and only for an instant, at her eyes. She gazes up at me, silent, hidden.

The eyes of wild animals are unfathomable pools. Most of what we see in them is ourselves, reflected.

I heard once about a study of scientists doing research on chimpanzees, I think it was. According to their culture, with amusing predictability, the scientists asked completely different questions about these animals' behavior. The Japanese wondered about teamwork and cooperation. The Americans examined dominance, territoriality, and aggression—certainly the main themes of many nature shows I see on television. Near Qinghai Lake, where countless birds gather each year to breed on the high and remote Tibetan plateau, I watched a pair of Chinese documentary films that showed how large masses of birds work together as a community of equals to repel outside predators—the Russian bear, no doubt, shape-shifting like a trickster into fox, weasel, snake, hawk.

For myself, I feel the allure of arcane detail about other creatures. Adaptation subtleties, like the way butterflies fold their wings into a shadowless wedge when they get too hot. Hints about how utterly different the world must look through multiple eyes, or eyes that see a different register of light. I like to know that in Nepal deer climb trees, while in Thailand they bark. That zebras bark, too. That in the fall American elk bugle with a sound like a rusty swing, starting low and ending a long time later on a high screech. That some enormous birds really are bright pink.

We can look to wild animals for the known and familiar or for the strange and unknown. A day or two after the films at Qinghai Lake, I visited a Tibetan monastery, an entrancing and exotic place. In the courtyard of one building, on the balconies of the second floor, stood lines of stuffed animals. A horse, a bear, a yak, a row of cowlike and deerlike creatures with oddly curving horns, flat and rounded faces, open mouths and staring eyes, all of them decked out in brightly dyed prayer scarves and clumps of yak hair glued on with yak butter. Until that moment I don't think I'd realized that there were large mammals I'd never heard of, never seen even in photos, and I didn't quite believe my eyes. I bought a picture book to help me remember them: in two languages, two scripts—Chinese and Tibetan. Were these creatures the constructions of humans, like the jackalopes you see in bars in small towns across the western United States? Were they mythic figures, the centaurs and unicorns of Asia? Emblems, like Sasquatch or the yeti, of what we might not know or of what we might wish were true? Or simply and amazingly real, the familiar neighbors of Tibetan nomads, as much a part of their home as bighorn sheep and bugling elk are of mine? How rich I felt, seeing these strange figures, these enigmatic bodies.

The best questions are the ones we can't answer—or the ones we don't even know how to ask.

The night the United States first bombs Baghdad, it is sunny, cold, and calm in Wyoming. John and I ski up the North Fork of the Shoshone River into the wilderness.

The tracks of animals are bright in new snow.

Moose and elk have walked before us, plunging across our path, up and down steep slopes hip deep in snow, leaving soft, round craters to mark their sleep. Somewhere near us they wait, graze on willow and aspen, nap on their feet in thick forest. Water ouzels dip in clear pools, flipping and splashing, vanishing when they hear us, reappearing if we wait quietly. All morning we follow a mysterious track—a long, delicate, gentle scoop wavering across the riverbed, tracing the fragile edge of ice, marking open water—and wonder what could move so serenely through the world. Just as we stop for lunch, at the edge of vision, for a slow instant, we see. Round, dark brown, big as a small bear, soft-edged, padding into the trees. Badger? Wolverine? Mystery.

At lunch, two ravens fly down the river, black on blue, bright sorties of wind and exuberance, twisting wing over wing. Thin flags of snow slip from high ridges. Around us great brown bears sleep, frost-grizzled, cradled in bunkers of ice. The hills shimmer, fine black lines filigreed over dazzling white, etched by the big Yellowstone fires.

Within days the Tigris and Euphrates will fill with blood. The air of the Fertile Crescent will darken with smoke, the Persian Gulf with spilled oil.

We turn toward home, retracing the twin strands of our ski tracks. Above us an eagle circles, white-hooded, watching us watching him.

John and the woman who has stopped to help hurry back to the fence with pliers. He bends close—like me, he tells me later, he barely glances at the doe—and studies the strands to see how they are woven and where they will come apart. He twists and pulls. In just seconds, it seems, he unwinds the last strand, the snare tying her to the fence. I open my hand and push her leg away from the wires.

She leaps up, stumbles, falls sideways into the fence, scrambles her way out, and lurches awkwardly away, trying to run with her leg still rigid and awry. "This happens to our horses," says the woman from New Harmony, calmly, "and they're always stiff for a while. But they get over it." The doe does seem to be loosening up, regaining her balance, as she disappears into the tall sage.

What is real, here? War and cruelty, hazard and suffering, our bombs falling on Baghdad, the sharp snares of barbed wires? Or dreaming buffalo and flashing hummingbirds, strange round animals disappearing down snowy rivers, a meadowlark singing on a post? None of it, perhaps, all things illusion. Or all of it: every life knotted together with others in an intricate tangle.

When I ask myself such questions, I think about something I read in one of Thich Nhat Hanh's books, *Thinking Peace*. He grew up in Vietnam during the war, surely a place to learn about suffering. "Our true home is in the present moment," he says. "The miracle is to walk on the green Earth in the present moment, to appreciate the peace and beauty that are available now. Peace is all around us—in the world and in nature—and within us—in our bodies and our spirits." For me, a glimpse of a wild animal is a brief flash of light, no easy solution to the big problems but a moment of comfort, a step into this surrounding peace.

In the middle of the night, when I wake from nightmares or John can't sleep, when all we can see in the dark room is fear and loss, I draw up wild images to hold and study, little nature mantras. "Go back to sleep," I'll murmur. "Remember those ptarmigan we saw up high in October? How they matched the snow and tundra? How they huddled, and the soft chucking noise they made?" Or, "Think about the day we skied near Yellowstone and the animal that might have been a wolverine. Imagine the cold, the emptiness, the quiet. Sleep."

A pile of white fur on a summer trail. A flash of ravens against snow. A bobcat at dusk, there and gone.

Fear

I'm often afraid in cities and almost never in wild places. Freeways and sirens and the local newscasts make me uneasy, not weather or wild animals. Not all my friends feel the same way, of course. "Are there snakes?" they'll ask, a tinge of panic in their voices. "Are there mountain lions? Did you read about that black bear that dragged a man out of his camper and ate him? What if we get struck by lightning? How long would it take us to get to a hospital? How far to a phone? It's so barren and desolate here! These mountains seem so forbidding, so *unfriendly*. What if we get lost? What if a blizzard comes? How close do you think that coyote is? Will it come bite us when we're asleep?" Please, I want to answer, be realistic. The riskiest part of this trip by far was the time in the car.

Still, once in a while fear catches me, too, shrinks my chest muscles and chops my breath, distorts my vision like a surrealist movie.

Sometimes no reaction could be more logical. One of the staples of wilderness narratives is the brush with death. Climbers get stuck halfway up precipices or fall off cliffs. Hunters and skiers encounter blizzards, get hypothermia, lose toes to frostbite. Rafts capsize. Hikers get lost and spend days wandering in circles, hoping for rescue. Cautious as I am, I've been caught above timberline in lightning a handful of terrifying times. In one sudden violent storm in Wyoming, John and I crouched and trembled under a low ceiling of wind-stunted shrubs and talked about whether to hold hands so we'd perish romantically together or move apart so if just one of us was hit the other could try CPR. You hope to keep your head in such tight spots. You hope to fend off panic. But if you didn't feel fear, there'd be something wrong with you. These are the moments your body confronts the body of the earth most starkly. You can't ignore it: you'll die sometime, and that time might be now.

Other fears are more complicated, less rational. Sometimes the power of suggestion starts the process: the more I watch the ten o'clock news, the more every stranger looks like a criminal. The more you read or hear about the dangers of the wild, the more you'll expect to encounter them. At these moments, we might say, your mind confronts the literature of the earth.

Some years ago John and I set off to explore to the north, through Wyoming, Montana, and the Canadian Rockies, planning to take a lot of short backpack trips. At our first stop, Yellowstone, we made a major mistake: we started reading signs and brochures about grizzly bears and collecting bear stories from park rangers. Within hours, our trip was irredeemably skewed. Phantom grizzlies stalked us everywhere.

We'd planned our first night out for the ominously named Beartooth Mountains. They're at the extreme edge of the territory used by Yellowstone's few grizzlies—only about 265 of them in the whole giant ecosystem. We bought bear bells and tied them to our bootlaces. We packed only food that took minimal cooking, left no juicy packaging, and, we hoped, wouldn't smell. We made so much noise walking that we saw no wild animals of any kind and hardly any birds. At dinner, John dropped a tiny bit of jelly on his jeans and thought, half-seriously, "I'm going to die." I discovered that I was in the suddenly dangerous habit, when cooking over a backpack stove, of wiping my hands on my thighs.

We both skipped brushing our teeth in case a bear might like the smell of Crest. At every step, John said, he expected a giant beast to jump out and attack him. Just what I often feel in a city, walking after dark.

We decided not to go backpacking in the Bob Marshall Wilderness after talking to a ranger whose job included dynamiting dead horses along its trails. "If you leave them there," he said, "they attract grizzlies. We use so much explosive that when the smoke clears you can't even find the horseshoes." At the entrance to our campground in Glacier National Park stood a sign with giant letters: CAUTION! CAMPERS HAVE BEEN KILLED BY GRIZZLIES IN THIS CAMPGROUND! I remembered reading about it when it happened, in fact, years and years ago, one of those newspaper stories I wish I'd never seen. "My God," the young girl screamed. "He's eating me!" We spent nearly every night of our trip in motel rooms.

We were way beyond reason. Even at Glacier, signs say that encounters are usually survivable and very rare, much rarer than car wrecks. Bears nearly always prefer to avoid people. In a lifetime in the West, I've seen only a few, never a grizzly. A couple of years after this long trip, a big cinnamon-colored bear loped across the road in front of our car—in southern Colorado, not very far from where the last known grizzly in the state had been killed not so long before. We pulled to a quick stop and without thinking twice leapt out to try to see him again. No fear then, not even commonsense caution. Grizzlies are high on our short list of animals we hope to see some day in the wild.

Why were we so spooked? I have no answer, just some theories.

I often think I'm a chameleon for other people's moods: if somebody I'm with is cranky or nervous or happy, so am I. Maybe I just caught John's fear. He thinks he was still reacting to a recent and serious melanoma scare: deadly danger, for a few years, seemed to him to be everywhere. Perhaps I wasn't truly afraid at all, not with the visceral emphasis of real terror, but just playing a kind of game, as when my girlhood friend Donna and I would scare ourselves reading Nancy Drew books in my basement bedroom, then run shrieking and giggling into the sunny backyard.

Maybe my fear was kin to my feeling about spiders. So what if most of them are tiny and harmless? The black widows in the window wells of that same bedroom weren't. Or to the common fear of airplanes. Sure they're safe, statistically, but numbers say nothing about how it feels to be so thoroughly out of control. On one horrible flight, I wrote farewell notes in the air—thinking in a triumph of illogic that I'd mail them if I survived. (Dear Mom and Dad, Thank you for everything. I've had a good life . . .) And on every landing I remember what an experienced-looking businessman in the next seat said to me at the end of another, smoother, flight: "Whew. Dodged the bullet again."

Or maybe it's simply that we *should* be afraid of predators larger than ourselves—an evolutionary practicality I've mostly managed to repress. Maybe I was sensing something about bears themselves, whatever it is that makes them so central to myth. Humanlike, but much more powerful than we, they remind us of facts we prefer to forget, that we don't control the whole creation, that death is inevitable, that the food chain doesn't stop with us.

When I'm impatient with the irrational trepidations of my friends, I try to remind myself of the grizzly trip. Logic doesn't explain everything, I tell myself. It can't touch wildness.

Stunned by Beauty

I wake feeling full of energy. So I pack a lunch, talk John into coming along, drive south and up from the cabin where we're staying to the end of the road, and start walking into the Big Blue Wilderness Area, ready to explore.

It's late July, high summer here not far below timberline, on one of those stunningly beautiful days when the sky and air are perfectly brilliant and clear, and all the flowers seem to be blooming at once. Along the small stream on our right are lush tall gardens of brookcress, drooping mountain bluebells, clumps of some kind of bright sunflower, startling fuchsia flames of Parry's primrose. Pale clouds of blue columbine billow among the lime-lichened boulders on our left, brightened here and there by scarlet paintbrush and tiny violets, and I can hear the high sharp chirp of a watchful pika. I step onto a big rock, take a deep breath, pull a little red tin from the watch pocket of my Levi's and smear some tangy Tiger Balm on my chapped lips, coat my face, ears, neck, arms with sunblock, drink from my old red canteen, and study the prospect.

On our right is a marsh, a dozen small streams the color of root beer, wide washes of green and yellow and purple. Beyond it rises a ridge, trees first, thick blue spruce, then a smooth curve above them to the sky. "Let's climb up there," one of us says, and we plunge off the trail, across the soggy lowland, and into the trees.

We find a narrow freshet and follow it uphill. I swing my arms out wide and run my hands over the new growth that tips each spruce branch, sage-colored and almost as soft as fir, so different from the wake-up prickles of the older needles. Every few steps we have to climb over fallen logs, some still firm, others like feather pillows under our weight. I walk along one, testing my balance, slip on a wet spot, recover. A woodpecker clatters, then a worried squirrel. Now and then we come to a miniature meadow where the sunlight is stronger. In one is a bed-sized pool of moss, a brilliant parakeet green, edged in white brookcress. It smells cool and wet, but only a tiny puddle at one end is still liquid. All the rest has turned to velvet, even smoother to my fingers than it looks. I want to lie down and put my face against it.

We've got ridge fever, though—we want to get up and see the view before the afternoon clouds build and lightning threatens—so I pause only long enough to marvel, catch my breath, and go on.

We climb out of the last trees, take a few steps into yet another bright meadow, and start to scan the licheny boulders and scree slopes above us, looking for a walkable way up.

"Stop," says John, "Elk! Three of them, way up there, see? No, five. Eight!"

We sit quickly so we won't spook them. I get out the binoculars. Four more, then more again. A calf! It's still spindly, a bit teetery, with faint light spots on its back. More! They keep materializing out of boulders and low bushes as though we were conjuring them up ourselves.

They move left across the rocks, running a few steps, stopping for a bite, making the distance look like nothing. As they jump a grassy wash, I count them: sixty in all, three bulls with trees on their heads, maybe twenty tiny calves.

Are they really heading for that steep snowfield? Yes! In a wide straggling line they mosey on to the snow, take a bite, look around. A few lie down. Two cows rise to their hind legs and face each other, heraldic. A calf walks under its mother, stands upright, cranes its neck in an oddly graceful curve, and suckles. Another calf lowers its head, springs off the snow with all four legs at once, pivots, does the whole thing again. In the bright noon sun their shadows are concise, the bulls' antlers echoed fleetingly on bright snow as they turn.

The air is full of their cries, sharp plaintive yelps, a little like the cries of hawks or gulls, faint but clear in the still air.

Slowly they wander up the snow and disappear over the ridge.

I breathe again, and my heart starts beating again, or so it seems. It's like the excitement of first being in love, this plunge into beauty, little nerves ready to tingle all over my skin, my chest and throat and head packed with shimmering bubbles of light and space and moving air. "What a wonderful show," we repeat to each other, our voices full of exclamation points. Like magic! What a marvelous day!

A sandwich and an orange later, we head uphill ourselves, aiming toward the spot where we first saw the elk, away from the place they disappeared.

Encore! The ridge is broad and nearly flat on top, the view almost a full circle and very long, and the elk are still in sight. Have they seen us, too? It looks like it—in a quick tumble they rush for the horizon, stop, each one, on the very edge, in full silhouette, then vanish. I sigh. And now they're back, all of them, dashing for another big snow patch, where they leap and run, in circles and wild arabesques, snow spraying from their hooves, their voices faint with distance. Not spooked but playing, a show not for us but for their own long July afternoon. They cry and dance and kick up their heels, and then, once more, they cross the skyline and are gone.

On days like this, beauty grabs you by the throat and shakes you breathless all day long.

You can hope for this gift, but you can't demand it. Most days we're separated by sheets of glass from what's around us. But sometimes, when you're somehow ready, or just in the right place and moment, the glass shatters and falls at your feet in piles of rhinestones, and you see everything differently. For hours or for seconds, you're locked in the present, and you are absolutely where you are.

In the Middle Ages and the Renaissance, philosophers knew this experience and called it the marvelous, called it wonder. It's like a "systole of the heart," said Albertus Magnus, a thirteenth-century saint: wonder stops your heart for an instant, takes your breath, shocks you into vivid awareness. Columbus's diaries are full of such moments—flocks of parrots, amazing trees, every island lovelier than the one before—for to be stunned by beauty is to discover a new world.

One bright December day we ski up Castle Creek. It's very cold. The shadows are sharp. Extraordinary colors tint the trunks of the aspen—pale yellowish greens, deep rusty reds, the softest ivories. Diamond dust covers the ground, more colors than the rainbow, gleaming shards, glittering, glowing, shimmering. Our skis shush in the snow. Then, somewhere not too far away, coyotes break into sound. They yip and bark and howl, exuberant, full of life, and yip and howl again.

The Elements

LIGHT

Western light, most days, is brilliant, palpable. It has body. Its particles hit your skin with soft, firm thuds, its waves push and tug you like a warm sea.

Midday in summer, at high altitudes, the light is laser-sharp, cutting razor edges for every grass blade, leaf, rock rim, every vivid petal, stamen, seed, every tiny sliver and disk of shadow. From every surface light flashes vivid, vibrant. It saturates every color, separates each shade, tone, texture.

When I was little, I discovered that when strands of my hair pull loose and blow in front of my eyes, they turn to filaments of colored light. Indoor blondes and reds and browns and whites become outdoor violets, greens, vermillions, golds. The colors of light. My own feathers, iridescent.

Backlit, when the sun is aslant, flowers and leaves catch the light like stained glass, hold it inside, glow.

Late afternoons and evenings, summer light turns red-gold, pools in meadows and sandy arroyos, washes over rocky peaks, stains the pinks of sandstones and granites a deep, soft, silky red. This is the Romantic light of Bierstadt's western paintings. Like the neon gaudiness of desert sunsets, it looks unreal on canvas or postcards, intensely real on the land.

If you stand just before it rains in a grove of aspens, the light will be green and liquid. In the fall, when the sun is lower and the leaves have changed color, the light, liquid still, will be gold. Through the shifting stencil of leaves, the sky's deep autumn blue will seem even deeper, more saturated. One September, after a watercolor class, I saw why—blue and orange, opposites on a color wheel, complementary colors, intensify each other. Pigment and light: circles of red and gold on a cobalt ground.

Winter light sharpens again, mirrors summer's brilliance and heat with a narrower palette of colors, blacks and whites, deep greens, bright blues. Variation comes in surprising places. A meadowful of willows will flash russet and saffron and claret. Up close, the white of aspen bark will reveal itself as pale green or burnt orange. Shapes etch themselves in your head, stay with you into sleep. On fresh snow, light fractures into diamonds. Tree shadows are sharp, but blue light gathers softly in snow hollows, rabbit prints, the round beds of elk and moose, ski tracks. At twilight, space fills with blue and violet.

The effects of light are countless. "Look at the light," I'm always insisting, pointing to a bright bar shooting across the sky, to the gleam of a spider's web high in a tree, to the sweep of silver on a field of grass, to the glimmering ripples on a streambed. When I travel I note differences shaped by latitude, altitude, humidity, measure them always against the quality of light that marks home.

Many places seem dim, their shapes and colors dulled by thick air, blurred skies. One fall when I lived in Ohio, the sun did not appear for forty days. The local newscasters kept count. To read at noon by a window, I needed a lamp. In places like this I feel dim too, heavy, sluggish, full of murky shadow. "What can you do," the postmistress said. "It's like this everywhere." *No, I wanted to shout, It isn't!* Spoiled by western light, I'm particular, a connoisseur, a snob.

I live on light so bright I have to keep checking to see that I have my sunglasses on; on light so intense that at evening I feel sunblasted, exhausted and scoured by light as if I'd been scrubbed by blown sand; on light that rushes like fast water into my eyes, my pupils small and black against the flood, my irises wide and dark, the color of sunlit storm clouds on a July afternoon.

ROCK

I know almost nothing about geology, but I love rocks. Smooth oval river rocks, painted like Easter eggs. Slabs of mica, layers of black translucence. Dark soft gray slate. Sandstone in pinks and reds and buffs, slightly rough against skin. Pink granite, every handful a hundred flat planes to catch the sun, a hundred rough edges to catch the shadow. A chunk of lapis lazuli from a secret mine, the midnight sky in my palm.

I've seen fossils of seashells and ferns on the top of a mountain nearly fourteen thousand feet high. I wander around old volcanoes, trace the errant paths of ash and lava, aa and pahoehoe, study chimneys and plugs, play with light puffs of pumice. At home I sit outside and try to grasp the story of the rock underneath me. It's very old, compacted red dust from an eroded mountain range, the Ancestral Rockies. I can barely begin to picture it, how they grew from the vanished traces of still earlier peaks until they were as tall as the Himalayas, then wore away into lush, dinosaury plains. The familiar Rockies rose later, just a short walk to the west, shoving the old plains out of their way into a fringe of hogbacks, narrow red ridges with gentle slopes on the east, cliffs on the west. Our house sits among these small ridges.

When Thoreau climbed Maine's Mount Katahdin, he found himself stunned by the raw power of exposed rock. He wrote about it with a mixture of terror and wonder, in language that reveals both the continuity in the concept of the marvelous and one fault-line in the concept of the sublime, one place where its structure has shifted in the years since. Mountaintops are hostile, he felt, no place for humans: "Only daring and insolent men, perchance, go there." But he also felt the powerful appeal of such places: "What is it to be admitted to a museum, to see a myriad of particular things, compared with being shown some star's surface, some hard matter in its home! I stand in awe of my body, this matter to which I am bound has become so strange to me. . . . Talk of mysteries!—Think of our life in nature,—daily to be shown matter, to come in contact with it,—rocks, trees, wind on our cheeks! the *solid* earth! the *actual* world! the *common sense! Contact! Contact! Who* are we? *where* are we?"

I climb another familiar ridge not far from Skyland, this one tall, among tall mountains, the trace of another massive event deep in the past. Along the top are level patches of fine gravel, purple, green, black. I choose purple and lie on my side with my head propped on my hand. Before me is space and rock, high peaks as far as I can see, white, gray, maroon, rose, several shades of each, shifting with moving shadow: the fascination of pure structure. The gravel beneath me is warmer than the air or the light. It presses hard against my hip, my ribs, my upper arm, sharp and solid. It writes on my skin. My body, earth body, ridgelines.

SPACE

Space, in the West, enters your body through eyes and the skin's pores, carried by light and wind, and flows through you like air, like blood, like the electricity of nerves.

For me it marks home. A dozen years in Virginia and Ohio, and I never felt quite comfortable: the space was the wrong kind. Clear around the world, in western China, in Mongolia, in Kenya, it was right, and I felt instantly at home.

Home space, huge and open, lets me see a long way. The air is dry and clear. The land can be flat or precipitous, I can stand on a mountaintop or at the bottom of a canyon, and still I feel the presence of space. Space is empty and full; it is not there, it is here.

A certain slant of light reveals how space is filled—with gleams of cobweb, insects like flecks of silver, golden dust motes, floating seeds, waves and particles from the sun's distant fire, the swoop and waver of swallows. Light pools in space, poured in and stirred by wind. At the ends of the day or in mist or with distance, space blues and blurs, turns soft between canyon walls. If you practice, you can learn to see space everywhere, between the edges of aspen leaves and pine needles, in the hollows of columbine spurs.

In the night sky in a dark place, space is insistent, intangible but substantial, invisible but embodied. A friend tells me how she once learned to lie still on a summer night, her back firmly held by gravity to rock, and perceive herself as looking *down* into the sky. I imagine floating on my stomach on the still surface of deep space.

"Breathing in," says a Buddhist exercise for mindfulness, "I see myself as space. Breathing out, I feel free."

WATER

In my life, water has had the character of punctuation. I amuse myself by pairing memories with punctuation marks.

An exclamation point: One evening we drive north toward Laramie. We're in shadow, but the valley is full of light. Suddenly water splashes on the windshield and we look up. A fine thick curtain of rain fills the sky to the west, each drop catching the sun as it falls straight to the ground, each drop a tiny shining sphere. Brightness falls from the air.

A question mark: It's an October night, but the water in the Gulf of Mexico is still warm enough to stand in. I'm shoulder-deep, unsteady in the push and tug of the waves, waving my limbs around like an anemone or a squid. I hold my arms out to my sides and pull them together. They glow pale green and flicker. I stand still and the water darkens. I lift one knee to my chest, straighten it, sweep a slow circle. Pale green, shimmer, darkness. I float loose of the sand, tread water, gild my whole body. I know the words—phosphorescent algae—but only the words.

A comma, tying sequences together: spring melt. Quick rivulets circle my tent, one every few feet all across this meadow, zipping down the hillsides to the icy creek just below, then, miles later, into the slow silty Colorado. At sunset black clouds drop torrents. Cocooned in blue nylon, I imagine the life of an ouzel, a water dipper, tucked in a moss nest on the edge of a cascade. Water rushes by like wind.

A dash—a pause, continuation: This island, Dominica, is one of the wettest places on earth, and the rainy season is starting right now. We eat breakfast on an open-sided terrace and gaze into a green tangle. While we sit, it rains five separate times, silently, air turning to water.

A semicolon, making connections between different but related things: If you stare at a waterfall, even a very small one, and then shift your eyes to the framing rock, the rock will move. Illusion; reality.

A colon, introducing what will follow: The lake at Skyland is still. I drift in a canoe, the soft waves of my own motion slapping lightly against aluminum, waxy yellow lilies and deep green leaf disks all around. The glassy surface doubles aspens, sage, mountains, sky. I lean over and trail my hand in the cold water. My face floats over clouds.

WIND

The wind will make you sick, they told me in China when I opened the window in the train. It will blow your body out of balance.

But I keep my balance against the wind. I lean into it and let it hold me, push me firmly upright. Wind blows sickness away from me, out of my head and lungs, scours my skin, empties that thick darkness between cells, fills it with cool, moving space.

The wind blows cloud shadows across rock fields and prairies. It whistles and roars and murmurs in my ears. With blades of grass it writes circles on desert sand. It sweeps over meadows in silver washes, turning to the sun the shining bottoms of grasses and willow leaves and aspens. In even the softest breeze, everything moves. The hairs on my arms sway lightly with spruce tops and reeds. Around my eyes and mouth and on my neck strands wander and whip and drift. The wind moves storms through, hail, lightning, snow coming hard and horizontal. It scrubs the earth clean, washes everything.

When the wind is soft or still, it fills with scents you can't describe but recognize with every cell. Dry sage and hot dust. The musty lilac of lupine at high altitudes in the sun. Pine sap. The metallic smell of rust-colored rock at old mine holes. Stock and wild roses. Ozone, summer rain, wet sage. When I breathe in, scent pools in my lungs, flows in my arteries.

I scramble through an explosion of sage and up a short cliff. I nestle into the rock where I'll be out of the strongest gusts, lean on a juniper, rest my head against its flat needles. Gazing at the slow green river, I dream, then drowse. A sudden tug at my ponytail wakes me up. Wind in the juniper, nudging me to attention.

LIGHT, ROCK, SPACE, WATER, WIND

I'm standing on the rim of the Black Canyon of the Gunnison. It's late afternoon, so bright that even with sunglasses I have to squint a bit to see. The wind blows hair into my eyes and presses my shirt against my skin. I lean into it for balance.

Under my feet and before me across a deep chasm of space is hard Precambrian rock, well over a billion years old, some of the oldest rock on the surface of the planet. It's nearly black, matte and mysterious in shadow, gleaming and brilliant in light, shot through everywhere with broad bands of pink. Two thousand feet below I can see white water, its speed and power faintly audible in a soft, insistent roar.

Just below me a small flock of birds materializes, made, it seems, from wind and space. Pinyon jays, a soft, light, slightly gray blue, robin-sized, flipping and turning in synchronized unity, a score or so, sunlight flashing shades of blue on their backs, the tops of their wings, the color of the sky at dawn, of water under light cloud, of washed silk, the palest of juniper berries. Water on wing, wind-tossed space-shards, light moving on rock.

Elated, alive, I stand on the prow of the planet, my face into the wind, arms flung out, flying through space.

Basking in the Sun

Winter. Even through dark glasses the sun makes us squint like night creatures thrown into day. We've shed every possible layer, stuffing our daypacks and wrapping our waists with fat sweater arms, but we're still sweating from the mix of solar heat and exertion from cross-country skiing. It's the last day in January. We climb just to the edge of the trees and search until we find an accidental bench—two logs wedged to make a seat and a back. With our ski poles we scrape them clear of snow. John pulls a pair of folded black garbage bags from his pack and spreads them neatly over the damp wood. We arrange ourselves, gingerly, shifting bones and flesh until we can feel no sharp spikes of wood, until we find the right angles of repose, add another layer of sunblock, and tip our cap bills low. My favorite time of day: basking time.

Spring, late May, floating slowly down the Green River, through a canyon misnamed Desolation. I'd call it Waxwing, maybe, or Canyon Wren, or Easy Swirl. We've propped the oars on the raft's sides and are lying back on our pile of gear, basking. I don't know when I've felt so relaxed, so exquisitely empty-minded, so dissolved into my surroundings, so *blissful*. Swallows circle and dip high over the cliffs. My sunlit eyelids are a glowing kaleidoscope.

We spin sleepily around with the river's hidden currents. The cliffs drift by in great slow arcs.

And summer. We've climbed high onto a ridge and we're feeling lazy. I find a narrow hollow between chunks of glittering granite and nestle until I match the contours of the ground. I prop my boots under my head and lay my binoculars on my stomach. When I wake up, I see not far to my side a wide slab stained neon orange with lichen. On it lies a soft pillow of brown, pointed nose raised just a bit for a better view. Watching light and shadow drift like lace across the landscape, we are mirror images—I on my back, he on his front. A marmot on a rock, basking in the sun.

These are sensory moments, simple animal pleasures. Under the sun's spell, I lie absolutely still and savor the sweet wild fruit of effortless meditation. Touch and hearing sharpen—the cool roughness of granite or tundra, the soft warmth or bright heat on my skin, the whisper, whistle, roar of the wind, the scratch and swish of grass blades rubbing together, the metallic buzz of flies. Time evaporates, drifts into imagination, and my body's indolent edges melt into ground and air.

My pulse must slow and my blood pressure drop. Some magic potion must flow out from the center of my brain, released by sunlight. At home, I read all the newspaper articles about the importance of sun to good health and spirits, nod as each new clue falls into place, my instincts confirmed by chemistry, physics, physiology, psychology.

Growing up in a dry, cool, sunny place, I didn't know my luck. Only later did I see the limits of geography and climate: in the rain forests of Thailand or the Caribbean, in the ovens of the Gobi or the Takla Makan, in the buggy taiga of Siberia or the soggy swamps of Florida, basking in the wilderness sun could be a pleasure only on the coolest days.

When the waters abated and the sky cleared, Noah and his wife and all the creatures of the ark must have chosen first to warm and dry themselves in the sun. That year in Ohio when the sun stayed under clouds for forty days, I huddled at home in our farmhouse with all the lamps on, hiding from my friends, all of us bored with my futile complaints. The next summer, John and I rented a tiny sun-filled cabin on the hot, dry Colorado Plateau and spent all but the most scorching hours outdoors. Oh, we covered ourselves with light cotton and caps and sunscreen and dark glasses—at least usually—but mostly we wanted to soak up the light, fill our pores with it as plants store chlorophyll.

We did not think of the sun as a danger. But that was the year that scientists first noticed the growing hole in the ozone layer over the South Pole—some five years after their instruments had started to record it. They were slow to recognize the problem, I've read, because everything in the past had taught them to expect anomalous data to mean broken instruments, not a broken world.

That winter, back in dark Ohio, I held John's hand while he recovered from anaesthesia: it had taken the surgeon four hours to cut the melanoma from his back, working out from its black widow center, looking through a microscope at each sliver of skin, hoping to find every cancerous cell. A tube drained blood from the six-inch seam, and stiff black wires poked aggressively through stained bandages. A long scar would remain to remind us of two stories—John's brush with death, the first crack of our sky breaking apart.

Fall now, November, in Arches National Park. We've scrambled up a few yards of smooth sandstone spillway into a high scoop of rock. The salmon walls rise straight up behind us and on both sides, and we're careful not to let our oranges or canteens roll down the slope at our feet. For a long time, we gaze lazily out at the window of view—dry wash, juniper and sage, a mesmerizing chaos of red rock, the mountains to the south already whitewashed with snow. An occasional lizard zips by, pausing to do a quick set of pushups. We move only to inch ourselves eastward, following the slow line of afternoon shadow. The rock beneath us is cool, the shadow chilly, but the late autumn light is warm. John keeps his bare skin carefully in the shade. "Remember," he says to me, "this is not the sun of your childhood."

I wrap myself in clothes and chemicals and tempt the fates: I lie basking in the sun.

Grubby

Consider the progression of the ages. Continental plates drift and collide, oceans rise and fall, inland seas advance and recede. Volcanoes spew out new earth. Mountains grow tall, then erode flat. Layers of sand and dust fill lake beds and ancient rivers, harden, wash away, settle again.

Chaos scientists call it "self-similarity across scale"—the way structures stay the same whether they're large or tiny. A cascade the size of your hand and Victoria Falls. The lines in a piece of driftwood, the contours of dunes, and the shapes of the Colorado Plateau seen from high above the land. Sedimentation and grubbiness.

Perhaps you begin with a layer of sunblock. Then you sweat. It's so dusty that every step you take sends eruptions of fine debris into the air. The mosquitoes in the trees are thick, so you add bug juice, the more toxic the better. It tastes awful. The wind kicks up more dust—into your hair, lungs, nose. It sticks to your chapped lips and your teeth. At lunch you dribble some jelly on your jeans, smear it in with your palm. Or peanut butter. Or melted chocolate. (Spilled food is always sticky.) You envision big bear tongues licking you clean. In the afternoon you add more layers of sunblock, sweat, dust. A few drops of rain make it to the ground and tiny rivulets of mud appear here and there on your arms, neck, cap, knees, then dry to arroyos. At dusk the mosquitoes swarm: more bug juice. Aluminum dust from your tent poles and greasy black soot from your cooking pot and backpack stove settle first on your hands, then transfer to your jeans. If you're a fire builder, you add splinters and sap, the fine black dust of decaying wood and the penetrating smell of smoke.

If you're lucky, your hair will accumulate more dust than oil, in an effect not unlike that of dry shampoo. No shine, no flexibility, but not slimy, not glued to your head with a tube of nature's Brylcreem. In either case you'll have hat hair—smashed flat on the crown, maybe a Woody Woodpecker flourish around the back.

Perhaps you try to clean up. The clothes, of course, are hopeless, and new ones get dirty so fast it's hardly worth changing. Face and hands, then. The stream is icy, and you don't want to pollute it with your soap. So you fill a small pot, maybe heat it up a bit if you're really energetic, squat on your heels, and attack a few layers with a bandanna and Dr. Bronner's magical eighteen-in-one peppermint soap—brush your teeth! your dentures! wash your hair! keep off bugs! 100 percent biodegradable! smells great! tingles! cleans and freshens from head to toe! OK! You'll feel cleaner for a little while. But nothing will get rid of the silt that's filled every tiny line in your hands.

Bedtime is the hour of erosion. Producing what geologists call unconformity, some layers disappear completely, erasing chunks of history. During the night, remaining sediments undergo metamorphosis, recombine and harden in place. With breakfast, deposition begins anew.

Different landscapes, of course, produce different earth forms. On the beach, fine sharp sand and sticky salt water interlayer with sunblock and lemonade. In the desert, sand, sunblock, and sweat dominate. If you died here, you'd become sandstone. In a rain forest, your skin turns liquid and slick, decaying leaves cling to your clothes, leeches crawl through your socks and suck on your ankles. Here you'd become coal.

Sometimes cataclysmic events occur, volcanoes, floods, earthquakes, mudslides. You slip pushing your canoe into the swamp and sink hip deep in rank black muck, then drift through a spider web so enormous that the individual strands tug at you as they stick to your humid, sweaty, bug-juiced skin. You slide down an abandoned coal chute, dyeing your clothes permanently gray, filling boots and pockets and scalp with black gravel and dust. Mysterious bug bites swell up to the size and color of squashed plums, and you smear them with cakey pink calamine lotion that hardens into the cracks of dried mudflats. A patch of wet brush leaves thin black whip stripes across your legs and torso and face. It rains hard and you ferment under your poncho, your soggy socks coagulating with your prune feet. You slip in a fresh cow pie.

In time, if you lie still on the ground, you will disappear from view. Like petrified wood, like a fossil in slate, your living pores will be filled with the matter of the planet, your shape preserved for the ages. Exquisite camouflage. At one with the earth.

Exotic

I wake hearing the echo of some strange sound and look out through the gauze of mosquito netting into the equatorial dawn. It's my first morning in Africa, at my brother David's house on the shore of Lake Victoria. John and I arrived at dark last night, and I have no idea what awaits me.

I lie still and listen, hoping to hear the sound again. Rustling leaves. A sweet bird song. Low voices. The static blare of David's shortwave radio tuned to the BBC morning news. My small niece Joy peeks in the door and then crawls under the netting to join me. A loud harsh cry outside: that's it! Har-har-har, it sounds like, a long sardonic laugh. "What's that?" I exclaim. "Is it a bird?" "Yes," Joy says, "Daddy says it's a long-necked beetle muncher," and she chortles in delight.

This unexpected bird's cry seems a kind of fanfare, introducing me to a new world. How *exotic*. What other startling things will I find here? How often will that word ring in my head?

Two weeks later I'm gaping in amazement. The guidebook said this out-of-the-way soda lake was nothing special, but we were offered a ride, so we came. And what a scene!

The landscape looks strangely familiar. The low dry hills dotted with bushes and flat-topped acacia trees are much like the juniper and sage hills at home, and so are the light, the space, the tawny colors, the emptiness. And the geysers that bubble, spurt, and steam all around me—these, too, I've seen, at Yellowstone.

But the giant pink birds! Thousands of them, greater and lesser flamingos, encircle the geysers and carpet the glassy lake. Some float like storybook swans. More stand in shallow water, on one stick leg or two, long necks straight up or crooked like fancy canes, looped around to rest along their backs, twisted and curved like new ferns, like paper clips, like letters in Arabic or Sanskrit. Most dip and groom calmly, dreamily, but a few pace aimlessly in tight agitated packs, beaks opening and closing and snapping at one another. Now and then a single strand of five or ten comes gliding past on spread wings, not far above the water, their necks and bodies stretched into perfect airfoils.

I hear an occasional soft squawk like a rubber sole on a polished wood floor, a flamingo's quack, I suppose. But mostly it sounds like there's a waterfall just out of sight. White noise. Pink noise.

And they are *pink*. Their wings and beaks are tipped with black, but their legs are fuchsia, and their feathers cover the spectrum from the palest pastel to cherry red, each bird several shades in itself, some close to white, some gaudier than the loudest lawn decoration. Even the water looks pink with their softened reflections. At the edge of the crowd we spot a quartet of scavengers, three African fish eagles—white-headed like American balds and a bit smaller—and a single gargantuan marabou stork, neck and skull bare like a vulture's, grotesque. Each one pokes casually at a pile of peppermint feathers around its feet. Just the right contrasting detail.

Beautiful. And, despite the echoes of home, definitely exotic. Beauty is a category I'm content to leave unexamined, but *exotic* intrigues me. What makes this seem like the right word for some sights and not others?

The question first occurred to me a few months before this trip, when John and I were driving along the Gulf Coast from Galveston to Tampa. Just over the border into Louisiana, in territory that should have seemed familiar after the five years I'd spent in Houston, we'd turned north to search for the wildlife refuge a friend had suggested we visit.

It was late when we arrived and a storm was moving in—slaty sky, chill uneasy wind, darkness and lightning to the west. We bundled up and walked on a long boardwalk out over the swamp and into another time.

Unfamiliar water birds floated around us. Squat black ones with scarlet turkey faces, slender dark bronze ones with the curved beaks of ibis, tall white ibis with pink beaks, red faces and legs. Little brown rabbits swam past looking quite comfortable in the water: swamp rabbits! Just at the edge of the reeds I spotted another pair of swimmers. Giant muskrats? Beavers without tails? I knew they must be nutrias: legendary (though perhaps falsely) for their viciousness, introduced from South America some decades back for their fur, long since become pests. When they noticed us, they vanished into the reeds.

We wandered on, then stopped to look at two men throwing rocks into the water. Their target was an alligator, a massive one, maybe fourteen feet long, as many yards away. It hissed back, loud, angry sounds from deep in its throat. Secretly thrilled to hear this amazing noise, I swallowed my indignation at the harassment and gazed through the binoculars. How primeval it looked, those ridges and bumps, those sharp teeth and protruding eyes, very like a dinosaur, a little scary, totally enigmatic. I'd seen gators before, but this one, glaring and hissing, seemed more vivid, alive, electric.

A moment later we understood. "It" became "she," guarding a dozen babies. Without binoculars they were invisible, but with them I could see clearly. Even then they looked amazingly small, so tiny they lay on the tops of floating grass blades. One astonishing sight—a monster from the primeval slime—turned into another: a mother protecting her babies, each one a perfect miniature.

I glanced up to check the approaching storm. Low sky and rising wind. Toward me, out of the bright steely gray, flew three enormous pink birds. They were so big, so close, so silent that I could hear the whoosh of their wings pushing down against the air. Three more, then another, maybe a score in all. Some passed us in profile—long white necks, thick pink bodies, stringy red legs. Others flew directly overhead showing off their six-foot wingspan. In the eerie storm light they glowed a soft rich rose. And then I noticed their beaks—long, flat, rounded out at the end into the shape of a giant spoon. Silly, like something invented by Dr. Seuss. Unlikely. But real. Roseate spoonbills.

It was a scene from an earlier world, when much of the earth was swamp, when mammals were small and reptiles large, when dinosaurs hatched their eggs and guarded their babies with bared teeth and ominous noises, when enormous flying creatures crossed the sky. *Exotic.*

Yes, but why?

The nutria would seem to be a relatively simple case. In a Louisiana swamp it is just plain out of its proper place, moved by human hands away from its home. It is alien, from outside: one of the meanings of exotic, relatively unambiguous. But value judgments, as always, add complications. Are nutrias "good" exotics or "bad"? Are all such exotics "bad," at least in places we want to consider wild? Alien kudzu and water hyacinths are notorious for driving out native plants and altering the character of wild places. So are tumbleweeds—despite their misleading place in western folklore, they came from Russia in the nineteenth century. Or tamarisk, a more recent introduction that's spreading along western waterways, choking out the native willow and cottonwood. Once I stumbled upon a menu in Green River, Utah, that offered an "old Indian myth" involving a Ute princess and a tamarisk—a weird attempt to twist one kind of exoticism into an ersatz version of another. But oxeye daisies are exotics, too, and I love them anyway.

What about the moose in Colorado? They're flourishing after being introduced where they'd never lived in any numbers, if at all. But my friend Geoff at the Division of Wildlife tells me that before hunters and settlement pushed them back, they were expanding their range in this direction, that left alone they'd have been here by now without any helping trucks or helicopters. Are they exotic? They certainly look like it to me—even in Yellowstone, where they are clearly native. Or how about horses? After all, miniature versions lived on this continent millennia before the Spanish reintroduced them. When we led our two llamas past a stray Hereford in the Wind Rivers, which of us all was the most exotic? The llama has been in this hemisphere, at least, for millions of years. Imagine coming upon a wooly mammoth or a saber-toothed tiger. What would we say then?

So what links the outsider-exotic to the evocative-exotic of my two encounters with enormous pink birds? As far as I know, other than the nutria, all the major players in those scenes are native species. Except John and myself, of course. But this is important. Often what strikes me as exotic is simply what I have not seen before—the unknown, the unexpected. Even pictures hadn't prepared me for the reality of these places, these animals and birds. I don't really think of elk, or even bighorn sheep, as exotic. I see them too often, too close to home. John's brother in Florida thinks nothing of the alligators out his back door, unless it's to worry that they might eat the neighborhood geese. When I first discovered white pelicans—every bit as enormous and quite as beautiful as the flamingos and spoonbills—I saw them as exotic. For several summers John and I went out of our way to find them and then watched them for hours. Now I've seen enough that I'd not use this word. Part of what it means is "unfamiliar," "surprising."

Perhaps the exotic also has to do with the juxtaposition of disparate things—as in surrealist paintings. Pinkness and largeness in a bird, big birds and geysers, bunnies that swim instead of hop, alligators and pink birds. Certainly such mixes add emphasis. But "disparate" is slippery. Would pinkness and smallness make a more ordinary bird? Would roseate spoonbills or alligators with babies seem more mundane paired with goldfish or cattle? Surrealists' combinations seem unsettling and dreamlike, even nightmarish, but those of the exotic have an intensified reality—for after all, they *are* real—and they command extraordinary attention.

Still, there is something about the exotic that is like art or literature. Words themselves can evoke it: Great Rift Valley, Lake Victoria, thorn trees, marabou storks, roseate spoonbills. Nursery rhymes, novels, movies: words and pictures create our sense of the world beyond home. For me, the quintessentially exotic landscape might be the home of Kipling's Elephant's Child, with his "satiable curtiosity"—the banks of the "great grey-green, greasy Limpopo River, all set about with fever-trees." It was the language that mattered, long before I learned to my delight that there really *is* a Limpopo River, and, more, that it flows through Mozambique, another alluring name.

Yet for me, neither the flamingo lake nor the spoonbill swamp had such clear literary associations. Still, somehow pink birds, alligators and crocodiles, Africa's Rift Valley, and Kipling all come together into a nebula of vague associations that can only be cultural. Surely they're traces of the time when Europeans began to explore other continents, far from their homes and unimaginably different—Africa, the American tropics, Asia. I can remember when even the word "native" suggested the exotic—"native dress," "native customs"—in a clear legacy of European exploration and empire. How strange that *native* and *exotic* could work as synonyms as well as antonyms! Thinking this I realize I've circled back to the paradox raised by the nutria and the alligator, one native and one not, both exotic.

All through my month in Africa I keep saying to myself, in a half-conscious refrain, "My brother's backyard is an exotic wilderness." On weekends, when things are quiet there, giant monitor lizards emerge from the dense shrubs and scavenge the compost heap, gnawing on bones and scraps of meat. Vermillion kingfishers and crowned cranes—nearby Uganda's national bird—fly overhead. On a tree in the far corner live some giant caterpillars my sister-in-law Nzali knows how to cook. Another tree offers bark that will cure nightmares. Venomous snakes might be anywhere.

Every morning at dawn John and I are awakened by the exotic, brassy cry of the long-necked beetle muncher, whose other names turn out to be equally appealing—hadada ibis and *Hagedashia hagedash*.

In the middle of our visit we leave to see other parts of Kenya. When we return, as we drive through the gate and into David's yard, a large bird passes overhead, laughing har-har-har. John and I look at each other and say with one voice, "We're home."

We've learned to feel at home in an alien place—surely a kind of gain. But I suspect we've lost a little, too. On the flight over, between Denver and London, I'd read a newspaper story about an Italian politician who had urged his audience to have more children rather than save their money for luxuries like travel. "The smile of your child," he'd said, "is worth a hundred times your desire to be free, to see Peru, Colorado or Burma." Peru and Burma stand high on my list of exotic places. Suppose I could learn to see my own Colorado home in the same way. What new things would I discover?

Early in this century a group of passionate artists in Russia claimed that the essence of art was to make the familiar seem strange. Perhaps this is also one of the roles of the exotic, to alter and sharpen our perceptions. In a kind of natural figure of speech, giant pink birds embody the marvelous, alligators the persistence of the ancient and the extreme brevity of human time. The exotic brings into focus the endless variety of the world and its irreducible strangeness.

Abundance

My brother David and I stand silent in a vast plain, knee deep in golden grass. Behind us and on our right rises the exotically named Esoit Oloolo escarpment, blue in the afternoon and hazy with the smoke from distant grass fires, probably set by Masai cattlemen. To our left thick trees mark the Mara River. Before us stretches sheer space, punctuated by scattered acacia trees, their branches trimmed high on their trunks. Here and there rises a low bluff or an eroded volcano cone.

This is the Masai Mara, part of east Africa's Great Rift Valley, Kenya's corner of the Serengeti. We're gazing south into Tanzania, and we're stunned, not just by this space and its spare beauty, but by the luck of our timing. It's nearly a month too early, but somehow we've arrived at just the same moment as the vanguard of the great wildebeest migration.

This morning, looking for a place to eat our lunch, we'd discovered dozens of dead animals in the river, bloated but mostly intact despite the crocodiles basking along the banks. The smell, perfume for the cartoonlike vultures hunched in a nearby tree, was rank enough to drive us away—despite the odd magnetism of those floating backs, those twisted necks and curved horns. (In a shocked voice, David had cried, "Look, a hippo with a wildebeest head in its mouth!"—and it was!) So we'd driven south, looking for one that was still alive.

And here they are in front of us, two or three thousand of them, resting, sleeping, grazing. Narrow masklike faces, dark like burnt rock, sharp curved horns, shaggy beards and manes, sloped backs, many with dark ribbons down their shoulders and ribs as if someone had just poured water along their spines. Many are young, shorter and thinner but already quick and sure on their feet. (Early this morning we'd startled a Cape buffalo trailing the membranes of afterbirth; her calf was still wet but already walking, though not very steadily. How quickly they must learn, here in the country of lions and hyenas!) Many of them look at us with a surprise and curiosity perhaps as sharp as our own.

This would be enough to see at once, but there is more. Maybe two hundred zebras stand among the wildebeests in small clumps. They shimmer in the sun, their patterns more elaborate and mesmerizing than I'd expected: stripes curving in separate sections to follow rump, belly, chest, neck, twisting into tangles on their faces, continuing up into their cropped manes. On some of them, the black stripes are distinctly reddish. Topis—tall spiral-horned antelope in burnished bronze—stand alone on scattered termite mounds, looking like the haughty animals on coats of arms. Delicate impalas and gazelles ring the margins, visible without binoculars mainly by their pale gold colors. I spot a couple of clumps of warthogs close to the ground, a dozen baboons, a lone gaunt marabou stork. And the final touch—with my binoculars I see a family of elephants, the smallest ones just higher than the grass, and a trio of star-splotched giraffes, their heads even with the flat tops of those solitary trees.

"You know," David says, "I've seen stuff like this on TV, on nature shows. These huge herds, shot with a really long lens, and in the background a bunch of elephants and giraffes. But I guess I thought they faked it." He's clearly never seen anything like this, even after nearly twenty years in the wilder parts of Africa, and neither, of course, have I.

We search the far hills to the south with our binoculars. Everywhere stretch long lines of black, fine and remarkably straight, looking just like the column of army ants we saw yesterday. Later in the afternoon we'll watch one of these lines cross the road in front of us—five minutes, ten, one animal at a time, single file, oblivious to our car. They run with soft, quick, regular thuds. Over a million and a half wildebeests moving north out of Tanzania toward us.

So many! I find it hard to believe. I struggle to absorb it, the sight and the sounds, this amazing vision of plenitude.

Abundance is visible everywhere, of course, not just in such spectacular places as this. Consider the blades of grass in a meadow, the trees on a forested hillside—examples so common as to be proverbial. Look up into the summer air when the light is aslant and you'll see shimmering specks of life floating higher than vision. Tiny spiders, I've read, drift around the peaks of the Himalayas. Or dig a spadeful of earth and contemplate its residents. Abundance is even plain in places much affected by humans—gardens, cornfields, tenements, our own interior organs. It's a reminder that some kind of wildness is everywhere.

Then why should this spectacle of wildebeests be such a thrill, and so moving?

Partly it must be the enormous numbers themselves. How many individuals does it take to make an abundance? Surely the answer is always relative. The fifty or so elephants we'll find on this journey seem like a lot to me—but then I've never before seen even one in the wild, and I've read so much about their recent slaughter that I'd have expected fewer. Not so long ago, herds of several hundred were not rare, and smaller groups like these were widespread. For my sister-in-law, one owl is too many; she's from Zambia, where (as in much of the world) owls are spirits of evil. For me, two spiders make an invasion. The hordes of grasshoppers in my garden two summers ago felt like a plague. So did the cockroaches we used to watch pour out of cracks in the kitchen walls when I was in college. Annie Dillard talks of such insect fecundity as a terrible reminder that the creation might well be senseless and mechanical. I suppose it would be logical to feel such a shudder here—but I don't.

And beauty? These wildebeests are far from beautiful. They remind me of the Billy Goats Gruff in my childhood book of fairy tales, funny-looking animals, almost ugly, somehow primitive, fit food for those dinosaurian crocodiles, though the longer I watch, the more appealing I'm finding them.

Beauty *was* important, though, in those masses of pink flamingos we'd seen the week before. And it saturated the abundance John and I stumbled across one recent January in New Mexico—the wintering grounds of nearly forty thousand snow geese and sandhill cranes. The black and white of the geese, the soft gray and red touches of the cranes, multiplied by thousands—those sheets of color, texture, and sound were stunningly beautiful. Or imagine the winter assembly in Mexico of the monarch butterflies—each one a jewel, so many together must look like lakes of topaz and amber and jet.

Sometimes, in cases like these, the allure of numbers is enhanced by the mystery of migration. Cranes, snow geese, monarchs, wildebeests—all travel great distances over ancient routes, and these large gatherings dramatize how little we know about how other animals find their way. (I wonder: Is it coincidence that sandhills and wildebeests both suggest the prehistoric? Could some of the biggest migrations be of the oldest creatures?) But if the flamingos migrate in any spectacular way, I don't know about it; they breed not all that far to the south.

A big part of what makes this scene of wildebeests so entrancing is all the *other* animals among them, each moving to a separate pattern of time and space. Giraffes, elephants, impalas, topis—the variety adds another level of pleasure. An abundance of one or two species is amazing. An abundance of many is something more, an overflowing chest of treasures.

"This is the kind of thing I like to read about in old African hunting narratives," David says. And I add, "In early American explorers' journals, too." We feel a kind of yearning for what we will never see, a nostalgia for the legendary and now only imagined experience of an undiminished wilderness, the landscapes and gatherings of animals of the past, part of the ideal we evoke with the word "pristine." Not so long ago the Great Plains looked like this, with carpets of bison, elk, deer, bighorn sheep, grizzlies, pronghorns, wolves, eagles. I'd thought the small buffalo herd in the Black Hills was probably as close as I'd ever come, so late in this century, to the sights recorded by earlier wilderness travelers.

But *this*, this is the thing itself. And it's encouraging to think we're just a short way from Olduvai Gorge, the site of the earliest known human remains. The landscape has certainly changed in all those millennia, grasslands advancing and retreating, volcanoes rising and wearing away, and I suppose it's likely that the wildebeests have not always followed this migratory path. Still, as long as humans have lived on earth, we have shared this ground with animals like these.

Mara, I'm told, means heart. So this Masai Mara is the heartland of the Masai people, as it was once for the human species. Abundance must be this heart's blood, the blood, perhaps, of all wild places. Sometimes with its promise of riches—the bison's warm coat, the elephant's lovely ivory—but very often just by its power to astonish, abundance has drawn us into the wilderness for a long, long time.

The Rare

On my dresser top lies a single butterfly wing.

It's only about an inch long. The top is coppery gold dappled with cream and edged with black. The bottom is pale sage green and silver—two dozen spots that shimmer like spilled mercury or illusions. It's a beautiful and delicate thing, so ephemeral that my finger tips can't even feel it, so light that if I breathe on it, it floats into the air.

Somehow, with the slanted and complex logic of the unconscious, this wistful fragment brings to my mind the intimacy between the abundant and the rare. If the top of the wing is abundance, sturdy and bright, then the underside is rarity—lovely too, but quite different, shot through, somehow, with loss. It might be more logical to say that abundance is the whole live butterfly floating in the sun, and that the single wing on my dresser is the rare—a partial remnant of what once was whole, a reminder of life now vanished, silvery as if in a mist, the slightly blurred luminescence of very old photographs.

I know my delight in abundance is tied to my sense of its loss. A kilometer from the spot where we watched a line of wildebeests cross the road is a fenced thicket, patrolled against poachers night and day by armed rangers, where there survived a single rhino. In *Moby-Dick*, Melville wrote about the inexhaustible multitudes of whales, how no amount of hunting could ever diminish their numbers. Under the wings of snow geese and flamingos are shadowy images of passenger pigeons, once so many they covered the sky for hours. Hidden among thousands of sandhill cranes is a tiny handful of whooping cranes, held tenuously on this side of extinction only by intensive care.

So I travel to wild places to revel in abundance, and also to contemplate, perhaps to see, the rare. Even in my own lifetime, I can't forget, the one can turn into the other.

Five centuries ago, when Columbus first reached the Caribbean, he was moved by the beauty of the land, the variety of trees, the abundance of birds. Dominica, I read somewhere, is the only island that he would still recognize, its dense rain forests largely intact, protected by very heavy rainfall and steep, difficult terrain. So John and I traveled there, wanting to see what those long-dead sailors might have seen, the home landscape of the Carib people. It was lush indeed, trees, vines, flowers everywhere, a river, we were told, for every day of the year. Columbus, amazed, wrote in his journal that parrots darkened the sky.

One day we hired a car and a guide, and late that afternoon, after a long, bumpy ride over a muddy road and a mile or so on a slippery dim trail, we watched the dusk fall at a canyon overlook. A handful of other tourists was there before us, but everyone sat quietly, gazing intently out over the impenetrable slopes before us, binoculars held ready.

Just before dark a pair of parrots glided silently through the dim liquid air, barely visible but somehow still resonant, bits of green forest tipped with scarlet and given wing. Of this kind there remain a few dozen birds. Of another kind of parrot, the enormous emerald and purple imperial, there are now only a handful. For these we waited past dark, in vain.

On the Mountain

When I went away to college, in one week's time I dropped from the cool silence of Skyland's nine thousand feet, where I'd spent the summer walking on mountains, to Houston's sultry, siren-loud sea level, and I immediately started looking for things to climb. A live oak by the library, where I spent hours reading and watching people pass underneath, hidden not by foliage but by the ground-gazing habit of walkers. The roof of my dorm, eight floors up, a bare concrete slab used mainly by sweltering sunbathers, where at night I could hear the roar of the lions in the zoo across the street. The library roof and the top of the astronomy building, locked doors and windows circumvented somehow in ways I can't remember, reached quietly at night with friends, the latter in October in hopes there'd be a usable telescope to see through the city's bright sky to the Orionid meteor shower.

And finally, late on a November night after physics lab, the most vertiginous point on campus: a construction crane set three stories deep in the ground and rising another three at least above. We started at ground level, crossing from solid earth over the enormous hole on a thin steel bridge, and then we took turns watching and climbing. My friend Jim went first but not very high. Then I climbed, all the way up, slowly, sticking close to the ladder, holding tightly with sweaty hands, weaving my arms through the rungs when I paused, testing each foothold before I shifted my weight. Under each step the crane swayed a bit, my stomach knotted, my heart quickened.

After that, I stopped pretending Houston's drained and paved-over swamps were like the Rockies. The top of a construction crane was not an adequate substitute for a mountaintop or a high ridge. It wasn't the adrenaline rush I was after when I climbed those peaks I was missing so much, but a collection of other rewards I'm still sorting out.

These mountains I'm talking about—they're not K2 and Denali and Kilimanjaro. Getting up one requires no porters or oxygen, no pitons or carabiners, no ropes or crampons or ice axes, no expensive permits or exotic visas, no headlamps for midnight starts. They're the everyday mountains you can just walk up with some time and energy and perseverance and a few pieces of simple equipment—jacket, rain poncho, hat, sunblock, water, a lunch, boots for ankle support. They need to be high, though, high enough to provide a really big view. Those old, soft, rounded, tree-covered mountains of, say, Virginia's Blue Ridge, those don't really count for me—evidence, I know, of my Colorado provincialism. You can't see far enough from the tops of those hills, and what you can see is mostly more trees.

Proper mountains need to climb above timberline. Out here these are common, several everywhere you look, rings of them, cirques, strings of peaks connected by high saddles, the occasional solitary giant, lots of colors and all shapes, the child's perfect triangle, flat-topped cliff-faced buttes, jagged spires and knife edges, hogbacks and palisades, huge rounded mounds, crazy quilts of miscellaneous mountain parts welded together inside the earth and then shoved upward into the thin high air to be cut and honed by glaciers. Every one of them is different, and on foot you learn their character.

Half my life has gone by since I first climbed this peak, the one I call Lupine for the intensely cobalt and sweet-scented flowers on the way down, one of the blue mountains I used to see across the lake at camp. And at least fifteen relatively sedentary years have passed since I last climbed it—years in which I've mainly been content with shorter climbs, to ridgelines, escarpments, high passes. It's a particularly alluring mountain, just under 13,500 feet high, standing alone above everything nearby, almost as tall as the 14,000-foot giants in the distance, some seven miles away across a map, several rugged days away by foot. Nearly the whole climb lies above timberline, too, most of it on loose, bare rock. We'd save this mountain for the very end of the summer when our biggest risk in trying it would be the weather, our muscles at their strongest, our blood thoroughly adjusted to the altitude.

I started thinking about going back up Lupine about a year ago, late last summer, idly at first, almost as a joke to myself about the gap between desire and reality—mediocre muscle tone and an allergy to working out, weak knees only partly improved by therapy, strengthening exercises abandoned some time ago, the energy and insouciance of twenty covered over by the moderation of thirty and the prudence of forty. Wouldn't it be good, I'd think, to climb Lupine again, a motive to get in better shape, an excuse to spend lots of time at high altitudes, just the right mountain to frame a big chunk of my life, a wonderful mountain, layered with translucencies of memory.

Slowly reality receded and desire took over. All winter I dreamed about being on the mountain.

I needed a companion, so I leaned on John until he agreed. Then my youngest brother, Bruce, stopped by with his family in Colorado for vacation, and on impulse I asked him, "Do you remember how to get up Lupine? Want to climb it with us a week from Sunday?" "Sure," he said. He'd been up four times to my three and thought he remembered the route. Conditioning? No problem, he took his tiny girls on walks around the Chicago parks. Knees? Bad, but we could go slowly, especially down hill. Altitude? Plenty of time. Desire? You bet.

Could we actually push ourselves to the top? I called a triathlete friend with a degree in nutrition and asked what to carry for a little extra help. Carbohydrates, she said. Forget protein and fat. No gorp, no Snickers, no cheese. Just bread, dried fruit, maybe one of those high-energy bars. Lots of water. And eat often, even if you aren't hungry. So I shopped.

And the weather? The driest year in a long time—lots of forest fires, not many wildflowers. But promising for us. Then the day after Bruce signed on, the summer monsoons hit the state, six weeks late, and we started watching the forecasts.

Though it sprinkled during the night, when we wake up the clouds are disappearing. Over breakfast I watch the sun hit the first ridge we'll climb, elation rising in me as the sky turns a bright, cloudless blue. Bruce arrives at about 7:30, and we stuff our packs, prepared, we hope, for anything. I squeeze all my gear into a fat waist pack and strap my bulky fleece jacket on top.

We're in no hurry. We cross the meadow and hop over a shallow creek on rocks; most summers this is gooey marsh. As we stroll up the valley Bruce and I talk about our route, planning to find our way by memory. The only trick is to avoid the rock faces and really steep scree slopes. An old mining cabin reminds us where to turn into the trees. We plunge into the dewy shadows and put our bodies to work.

We talk about knees. Bruce, who is a physicist and mechanical engineer, demonstrates torque for me, bending his leg at odd angles, holding his daypack straight out from his shoulder, then with his elbow sharply bent, arguing for short quick steps rather than the long slow ones I'm used to taking. We're both feeling stronger than we expected—it's the energy of excitement.

In about an hour we're at the top of the trees in a thin zone of sharply sloping meadow. More flowers than I'd expected are still blooming, but most have gone early to seed. I spot a bright green moss pond. A chunk of volcanic rock. A slab of slate with a layer of quartz-crystal teeth. A fresh elk print.

The ground becomes rock and steepens again, and suddenly

the climb feels more like work. My knees start to twinge every few steps.

We stop to rest and think about our route. Two does move quietly across the slope above us, startled perhaps by Bruce, who's checking out the slope to our right. Then a crash, repeated, like breaking china, and a fawn, its spots nearly gone, bounds straight down the scree toward us and past, twelve or fifteen feet at a leap, a quick almost-stumble, a longer pause in the air to rebalance, feet in loose rock again, another leap, gone before I can catch my breath. Wild.

Another familiar old mining cabin tells us we're on track, we reach the ridge, and the view to the north opens all at once. Small valleys and big mountains as far as we can see, a wide vein of marble, a handful of persistent snow patches. A few small cumulus clouds cling to the horizon. It's just midmorning, rain's not likely before midafternoon, and the rest of the climb is all on spectacular ridgelines. I feel like I'm full of bubbles, transparent spheres of iridescence floating me up into the sky.

For a while the ridge is broad and relatively level, the rocks small enough to make easy footing. Two steps to every breath. I remember a friend from camp telling me that in such places he always wanted to ingest everything in view. I try to look at everything all at once. What I see is both simple and complicated—just space, light, and rock, but such a vast and figured space, so many subtleties of light, such intricacies of rock!

The ridge narrows and steepens. We're at about 12,600 feet. Step, breathe in. Step, breathe out. It's razor thin now, 3 feet wide or less, the rocks on top flattened slightly into a kind of trail, space falling sharply away on both sides. When the path veers a step to the right, the rocks on my left side are suddenly shoulder high. Breathing in, I see myself as space. Breathing out, I feel free. A single columbine, tucked in the rock—a rare sight so late and so high. Earlier this summer there'd been an abundance of them, whole hillsides full. So delicate, so rugged. A sudden roar, and we all stop. Thunder? Jet. But those distant clouds are growing and darkening. We speed up.

Suddenly it's much steeper. Now I'm leading. I step, breathe in, breathe out, step, stop to gulp some orange juice. The rocks are bigger and sharper. I put my gloves on, lean over to balance myself with my hands, test each step for stability. We measure our progress by the view. That's Frigid Air Pass, way down there. We're higher than the rim of Hummingbird Basin, the mountain across from camp, the mountain behind it. Soon only the fourteeners are above us. There's the lake at camp! Gleaming, so far away that none of the new buildings around it show—this view is the same as it was half my life ago. Home.

I know now I can make it to the top. But all the rewards are here in this moment. I'm stunned by beauty, breathless with it, heart-stopped, afloat on elation. Bubbles pop and reform inside my head, my throat, my chest. For a magical instant, the world around me seems exotic, marvelous.

A deep rumble startles us to a stop. Big storm cloud to the north, wrapping two huge peaks in a dark sheath. We watch for a minute, see a strike, think the cloud is moving sideways, check that the breeze on us still comes from the sunnier south. But other clouds are building fast, several hours earlier than we'd expected, and we can't see to the northwest along Lupine's ridge, the most common direction for weather. We couldn't go down here even if we wanted to—it's way too precipitous—and the route back would be slow and very hard on our knees. A half-hour more and we'll be at what Bruce calls the Whale Back, a broad slope down to the south, covered with small, loose scree, good ground for a quick descent. We take off uphill, hoping the clouds will dissipate. John's just about decided to skip the peak, but Bruce and I are optimistic.

As the Whale Back comes into view, we talk about the way down—if we have lots of time, we'll follow that ridge all the way to the left; if we have to hurry, we'll go to the third snow patch and drop over the side. When we reach the point of decision, we stop, unpack sandwiches, and spread out to watch the speed and direction of the clouds. I drink from my red canteen, put my gloves away, switch from sunglasses to plain ones. Once, here, a golden eagle swooped across the ridge not twenty feet in front of me and just below. It's at least another hour to the top and back to this point—another four hundred feet up, half a mile over, one more knife edge. I'd dreamed for months of sitting on top, drinking in the world, looking for ocean fossils in the palm-sized slabs of slate, basking in the sun. Reality and desire wrestle.

Desire loses. The sky above us is darkening quickly. We'll get wet, we think, even though there's no lightning nearby, and we won't be able to see much. We snap some quick photos: we made it this far, we will say, we could have done the rest, but look at those clouds! We step reluctantly toward our packs.

Crack! Out of the silence a lightning bolt, too close to think about, and then hail and rain. We grab the packs at a quick lope and start down, shouting directions to each other: Aim for the lighter patches of rock, they're softer, more recently disturbed. Crack! Land on your heels and you'll slide better. Don't twist an ankle or knee. Crack! If it gets really bad, dive over the left edge. John disappears in front of us. I turn and pause every few seconds to check on Bruce, a lone, stiff-kneed figure behind me, silhouetted against rock and charcoal sky. I review what I know about CPR, break into a run.

Bang! This one's so close I plunge off the edge and hit the steep slope with my seat, propel myself down with arms and legs, flailing like a frantic crab. John's below me. Will he come back up if I get hit? Bang! Bruce sees lightning hit the ridge below him, right where I'd just been, and dives off the side onto a snow patch, slides down until he hits rock. Once I know he's off the top I stop looking back, don't see the bolts, don't count for distance, just plummet down the side, cutting and bruising my hands, jamming knees, pure adrenaline, terror.

Maybe eight hundred feet down we find a little ravine and stop to regather. I pull my sunglasses from my pack, stick them in the pocket half-way down my thigh, and sit on the pack hoping it will provide some insulation against electricity. My knees are shaking violently, my glasses fogged almost completely over, my cap bill pulled low and dripping. I put on my poncho. I count between flash and boom: one one-thousand, two one-thousand, three! I scrunch lower. Bruce announces that the bolts are hitting everywhere around us, mostly on the ridge top, at least once below us. I squeeze my eyes shut, feel guilty about bringing us all here, about dawdling up higher while the clouds gathered.

Should we stay here, wait it out? There's zero wind, a bad sign, and we're clearly in the middle of this storm system. The gully feels a bit safer, but we're still high, a thousand feet more down to the first trees, and the rocks are wetter by the minute, bigger and sharper, compacted by the rain, the footing harder, especially for Bruce and me. The ravine we're in is getting ominously deeper. Can we stay in it and keep going? Bruce walks over to see past the next bend: impassable. To move we'll have to climb back up on a small rib or two, feel even more exposed. I don't feel safe here, but the idea of moving tightens the knot in my stomach.

I try to remember every bit of lightning lore I've ever heard—and can't help remembering every story I've heard of people getting hit. The ones who survive unharmed. The ones who are stunned, then recover. The severe burns and long-term damage cases. The ones who die. Caves are dangerous, since the ground itself conducts. Would a ravine do the same? Should I try lying flat? But there was that man who was struck by lightning while he napped in his tent at the bottom of a valley. Should I keep sitting on my pack or squat on my boot soles? The usual folklore says soles, but my knees disagree, and only last week I read about some people who were badly burned after being struck in their *car*—a place I'd always thought completely safe. If tires don't help, shoes surely won't either. My niece Léa told me she'd been instructed to take off all her jewelry and throw it away from her. Right, I'd thought, the quintessentially useless action. A joke about helplessness. I check to make sure I'm not wearing earrings.

Now that there's time to think, I'm sure we're going to be hit. There's nothing we can do except hope and crunch ourselves into tight balls. The strikes continue, every half-minute or so, no farther away. Interminably. I'm not nervous or uncomfortable or apprehensive or worried; what I feel is plain *fear*. "Don't worry, Sis. Our odds are pretty good," Bruce says, and stands up until John barks at him to get back down.

John and Bruce decide we have to drop lower. I'm not about to stay here alone, so I hunch over and run with them across the rib into the next ravine, a shallower one, wincing with every knee bend, cringing with each slap of thunder, and turn to slide downhill. Here, near the water, just a foot wide but white with its quick fall, the rock is bare and slick. My heel slips, I slide several feet down, Bruce grabs and misses, I land on my pack, elbow, and palm, slide flat on my back into the water. The solid earth! Contact! Contact! I cradle my elbow and keep going, straight down the flow until I see enough loose gravel to hold my feet.

At last the lightning slows, moves a bit farther away. We can count to five, then six. Numbers that ordinarily would raise our blood pressure now signal reprieve. We reach the first willows, the ground flattens, the adrenaline slows, and Bruce and I can really feel our knees. We both start limping, trying to conserve what muscle and cartilage are left. I try to brush the gravel out of the skin on my hands. With every step I can see water bubbling out the toes of my new boots. I glance at a small pond beneath a tiny cascade and giggle at the thought that I'm carrying my own pair of pools. We check the time and can't believe it: it's been over two hours since the first bolt. No wonder it felt so endless! We swallow an aspirin each and start joking. "Good thing we had John along for emergency equipment." "Yeah, if I hadn't said we should go down, you two would be lying up there like charred chicken parts." "Lucky nobody sprained an ankle," I say, then twist mine, fall, and laugh. "Hey, we could try again tomorrow."

Finally we reach the road, barely paying attention to what's around us. I don't even notice whether any of the lupine are still in bloom or the late gentians have appeared. We take turns exclaiming: "It's almost flat! It's stopped raining! We're still alive!" And we trade stories—other brushes with lightning, hair standing on end, the smell of electricity, a friend of Bruce's who could see an aura around his ice ax, my next-door neighbor who watched a bolt shatter a nearby boulder, sending rock splinters everywhere. Our own story begins to take shape, each of us contributing details, sensations, making sense out of chaos, taming our fear. We'll keep telling it, nonstop for hours, frequently for days, every now and then for years.

Back at our tent we bundle everything together and throw it in the car. Not being masochists, we're heading for a hotel—drink, shower, dinner, hot tub. An hour later, when I start peeling off the sopped layers, black gravel sprays from my boots. My socks are dark gray—both pairs—and so are my feet. My pockets are full of soggy Kleenex and rocks. My pants and shirt are black, too, and my underwear. No blood on my palms, just scrapes, but I have to pick up my leg with my hands to get into the tub. I've brought the mountain down with me.

The next day we head home, delighted to be stiff and sore. I'm ready to retreat for a while, huddle among soft pillows. As he drives off, Bruce says, "That was fun. Let's climb another one next summer." I don't think he's joking. Within a week, I'm thinking the same thing, dreaming about being back up on the mountain.

Coming Down

The conversation is always the same. It'll be nice to be clean, Nina will sigh. Yes, I'll answer, a shower, wonderful. A cold beer, John will add, a burger, real coffee. Hot chocolate, a soft bed, someone might say, fresh clothes. And we'll start telling ourselves about our journey. Everyone will join in, rehearsing, editing and polishing the descriptions, the drama, layering these new experiences around and over our stories of other trips.

Every coming down is a palimpsest. Stepping carefully on a narrow desert path as dusk deepens, driving away from camp, Chaco Canyon, the Wind Rivers, the Masai Mara, skiing down a shadowy track toward the car on a cold February afternoon, wishing I could make the slow green river run even more slowly as the boat drifts nearer our take-out point, I remember all the other times I've left wild places.

Coming down, I feel the weight of my own body. I'm sunwashed and windscoured, my lungs and heart have grown stronger, my toes jam forward into my boot tips, my knees twinge, my hips ache, and my soles burn. I fall with the river toward some powerful force, and I can't slow down.

An empty space opens inside me, and I feel the coming ache of homesickness. When can I come back? Next month? Next summer? Will it still be wild? Where next? What's over that ridge? Does Alaska look like this gravelly streambed? When can I go *there?* I imagine other journeys, other wild places.

I remember Thoreau's "Talk of mysteries!—Think of our life in nature,—daily to be shown matter, to come in contact with it,—rocks, trees, wind on our cheeks! the *solid* earth! the *actual* world!" And the title of a poem by Richard Wilbur: "Love Calls Us to the Things of This World."

I think of beauty and transience, wildness and home, loss and desire. The inward arch of a glacier lily, the heady scent of wet sage, high mountains mirrored on the lake at camp, an incandescent circle of white pelicans against a gentian sky—like gravity, like magnetism, such heart-stopping images hold me fast to the earth.

September, a month after my climb and stormy descent, and I've returned to Lupine Mountain. My mother and I have driven here to see what this place looks like when the aspen are gold, when the bare willow branches glow russet and saffron. I walk a short way up the basin John and Bruce and I had rushed down, so filled with stories and relief that we'd taken no time to recognize the summer's end.

My mother waits for me in the car, too lame to hike just a month before her second knee replacement. Maybe next summer I can bring her here on a walk, I think, show her the sweet lupine and hidden waterfalls I've loved for so many years, until I remember she and my father walked in these mountains before I was born. What is she thinking now, I wonder, alone in the quiet valley as winter clouds slip like dreams across the vivid sky? I imagine her pulling a few leaves from the sprig of sage on the dashboard, rubbing them between her fingers, over her wrists.

I breathe in autumn, crisp and musty, and climb just high enough to see the Whale Back ridge far above me. Its three snow patches are a long cornice now, a line of shimmer between rock and air. As a red-tailed hawk circles low over my head, it starts to snow, quick pebbles of white thudding softly against my skin, and for a moment I stand still with my face to the weather. Seedpods bristle and droop at my feet, their crisp containers open now to the wind. I shake tiny pellets from a glacier-lily rattle, feel the stiffening velvet on the split crescents of lupine pods, gather a handful of seeds to take home.

Notes

I could not possibly mention the many books I've read that have affected this one: it's been too long in the writing. These are just the few I quote or paraphrase from most directly, ordered according to the chapter where they appear.

PREFACE

Roland Barthes's book is *A Lover's Discourse: Fragments*, trans. Richard Howard (New York: Hill and Wang, 1978). He explains his use of the term "figures" on pp. 3–9.

DESIRE

I found the passage from Sir Thomas Browne's *Religio Medici* in Dorothy L. Sayers, *Gaudy Night* (1936; New York: Avon, 1968), p. 252.

EQUIPMENT

On the paradox of simple living, see Colin Fletcher, *The Man Who Walked through Time* (New York: Random House, 1967), p. 28.

On washing dishes, see Thich Nhat Hanh, *The Miracle of Mindfulness: A Manual on Meditation*, trans. Mobi Ho (1975; Boston: Beacon Press, 1987), p. 4.

TRUDGING

For Margaret Murie's account, see *Two in the Far North* (1957; Anchorage: Alaska Northwest Publishing Company, 1978), pp. 146–47.

For Robert Marshall's, see *Alaska Wilderness: Exploring the Central Brooks Range* (Berkeley: University of California Press, 1970), p. 145.

GRANDEUR

For Thoreau's remark about the sublimity of waterfalls, see *The Maine Woods*, ed. Joseph J. Moldenhauer (Princeton: Princeton University Press, 1972), p. 75.

I've used the version of "God's Grandeur" from *Gerard Manley Hopkins, Poems and Prose*, ed. W. H. Gardner (New York and London: Penguin, 1953, 1958), p. 27.

WILD ANIMALS

This passage by Thich Nhat Hanh is from *Touching Peace: Practicing the Art of Mindful Living* (Berkeley: Parallax Press, 1992), p. 1.

STUNNED BY BEAUTY

I learned about the medieval idea of the marvelous and found the quotation from Albertus Magnus in Stephen Greenblatt's *Marvelous Possessions: The Wonder of the New World* (Chicago: University of Chicago Press, 1991), p. 16.

THE ELEMENTS

The two Thoreau passages I quote are from *The Maine Woods*, pp. 65 (mountain tops), 71 ("Contact! Contact!").

The Buddhist exercise for mindfulness is from Thich Nhat Hanh's *Touching Peace*, p. 12.

BASKING IN THE SUN

I found the date of the discovery of the ozone hole in Bill McKibben's *The End of Nature* (New York: Anchor, 1989), and have just slightly paraphrased his remark about the scientists' assumption that anomalous data meant broken instruments, not a broken world, p. 42.

EXOTIC

Rudyard Kipling's story "The Elephant's Child" is in his *Just So Stories*. I checked my memory of the line about the Limpopo River in the 1978 Weathervane Books edition (New York, 1978), p. 51.

Acknowledgments

I've had a lot of fun writing this book, and much of it I owe to my family and friends. Many of them have helped, and in many different ways, sometimes without knowing it—recommending places to go, telling me stories, giving me ideas, talking to me and letting me talk to them, reading and commenting on versions of my manuscript, encouraging me, and, most crucially, keeping me company in wild places. Many of these friends make at least cameo appearances in these pages, but some stay behind the scenes.

For all these things, I want to thank my mother and father, Nancy and Laird Campbell, who raised me to write a book like this; my brothers David, Douglas, and Bruce and their families; and my friends Bill Alldredge, Ben Brown, Russell Burrows, Brock Dethier, Mary Lea Dodd, George Finnell, Bruce Golden, Melody Graulich, Mark Jenkins, Susan Laidlaw, Don Scheese, Geoff Tischbein, Scott Thomas, Larry Wiland, and Lisa Williamson. By figuring out how to rescue my text from the isolation of WordStar and my museum-quality Kaypro computer, Gene Dodd saved me a lot of trouble and helped me out of the Middle Computer Ages. And Michael and Valerie Cohen gave me particularly useful readings of my manuscript and lots of ideas for improving it, some of which I'm afraid were beyond my reach.

Thanks, too, go to the English Department and the College of Liberal Arts Professional Development Fund of Colorado State University for their help in paying for some of my "field research." A sabbatical leave from teaching gave me the time to finish what was becoming a very long-term project.

I am especially grateful to four other people. Audrey Dorsett, who ran the camp where I spent many summers, showed me how to make myself at home in the mountains. My good friends Nina Bjornsson and Mary Golden have talked to me, listened to me, gone camping with me, read my manuscript more than once, kept encouraging me to write more, and made useful and challenging suggestions. And my husband, John Calderazzo, has been essential in ways only hinted at by his presence on so many pages of this book.

About the author

SueEllen Campbell teaches in the English Department at Colorado State University in Fort Collins, Colorado. Before attending Rice University and getting her M.A. and Ph.D. at the University of Virginia, she grew up in Denver and explored the Colorado Rockies. She has also lived in Wyoming, Ohio, and China, and has traveled extensively, both on foot and by other means. She has written widely about twentieth-century literature and American nature and environmental writing.